Passive and Active Microwave Circuits

Passive and Active Microwave Circuits

J. HELSZAJN

Department of Electrical and Electronic Engineering
Heriot-Watt University
Edinburgh, United Kingdom

A WILEY-INTERSCIENCE PUBLICATION

JOHN WILEY & SONS

New York • Chichester • Brisbane • Toronto

Library of Congress Cataloging in Publication Data:

Helszajn, Joseph.
Passive and active microwave circuits.

"A Wiley-Interscience publication."
Includes index.
1. Microwave circuits. 2. Microwave devices.
I. Title.

TK7876.H43 621.381'32 78-5787
ISBN 0-471-04292-7

Printed in the United States of America

10 9 8 7 6 5 4 3 2

IN MEMORY OF
MY FATHER

Preface

This book was written primarily as an introductory text in microwave engineering for final undergraduate or first-year graduate students in electrical engineering. It is also likely to be of value to engineers in industry. Since it is component rather than system oriented, it incorporates examples of the most important devices used in microwave engineering. An effort has been made throughout to maintain the length and standard of each chapter as uniform as possible. Except for the chapter on scattering matrices, which is required reading, each chapter is fairly self-contained. No background other than an introduction to transmission line theory and waveguide fields is necessary for understanding the text. Although the material included in the text is too large to be covered in one semester, it will allow each professor to construct a mix of topics without too much restriction and will also allow some latitude for varying curriculum from year to year. It is hoped that the student will find the text sufficiently interesting to follow up the material not covered in the classroom.

The three classes of devices treated in this text are passive components, nonreciprocal ferrite devices, and semiconductor circuits. The classic electronic tubes such as the klystron, magnetron, and traveling wave tube have been omitted since their descriptions are readily available in many standard textbooks. The book starts with scattering and immittance matrices, the former being essential reading. It continues with passive networks such as directional couplers, impedance and mode transducers, phase and attenuation networks, resonators, and filters in Chapters 3–7, respectively. The classic nonreciprocal ferrite devices, gyrator circuits, and circulators are described next in Chapters 8–10. The semiconductor class of devices is studied in Chapters 11–15, which deal with variable-resistance and -capacitance devices, negative-resistance bulk devices, nonlinear resistive mixer circuits, and field-effect transistor circuits.

At microwave frequencies the measured quantities of both passive and active circuits are most often the scattering parameters of the device. These parameters describe transmission and reflection at the different ports of the device. The scattering coefficients are therefore used as far as possible to characterize the behavior of the devices dealt with in the text. For the symmetric passive devices, the entries of the scattering matrix are formed by constructing the eigenvalue problem used in the classic book by Montgomery, Dicke, and Purcell *Principles of Microwave Circuits* (McGraw-Hill, New York, 1948). This approach is also utilized to formulate the nonreciprocal ferrite devices such as the circulator and gyrator circuits. Although the eigenvalue problem is not extended to the class of semiconductor devices, the scattering variables are still the measured quantities there, no more so than in the case of the microwave transistor amplifier.

Since this is essentially a teaching rather than a research text, no effort has been made to acknowledge individual contributions specifically, but it goes without saying that this work is but a reflection of many individual efforts over the past 50 years.

Wholehearted thanks are due to Sheila Murray, Moira Tullis, and Helen Vaughan of the Department of Electrical Engineering, Heriot-Watt University, for their good will and cheer, without which this task would not have taken root.

J. HELSZAJN

Edinburgh, United Kingdom
April 1978

Contents

Passive and Active Microwave Circuits

CHAPTER ONE

The Scattering Matrix

The scattering matrix dealt with in this chapter is admirably suited for the description of a large class of passive microwave components and is used as much as possible throughout this text. In many cases it leads to a complete understanding of the microwave device while avoiding the need to construct a formal electromagnetic boundary-value problem for the structure.

The entries of the scattering matrix of an m-port junction are a set of quantities that relate incident and reflected waves at the ports of the junction. It describes the performance of a network under any specified terminating conditions. The coefficients along the main diagonal of the scattering matrix are reflection coefficients, whereas those along the off-diagonal are transmission coefficients. A scattering matrix exists for every linear, passive, and time-invariant network. It is possible to deduce important general properties of junctions containing a number of ports by invoking such properties of the junction, as symmetry reciprocity, and power conservation.

Since the entries of the $\bar{S}$, $\bar{Z}$, or $\bar{Y}$ matrices of a symmetrical network are linear combinations of the circuit eigenvalues, their direct evaluation or measurement provides an alternative formulation of network parameters. The m eigenvalues of a symmetrical m-port junction are 1-port reflection coefficients or immittances at any port of the junction corresponding to the m eigenvectors of the device. These eigenvectors are the m possible ways that it is possible to excite the junction and are determined by its symmetry only. The 1-port circuits formed in this way are known as the eigennetworks of the network. In the case of symmetrical 2-port networks the eigenvalues may be obtained from measurement or by calculation by applying in-phase or out-of-phase eigenvectors at the ports of the network.

The scattering parameters of symmetrical 2-port networks can be readily obtained from their equivalent circuits by forming their eigennetworks. These eigennetworks are obtained by bisecting the 2-port network and opencircuiting and shortcircuiting the exposed terminals. The reflection coefficients of these two 1-port eigennetworks are just the two eigenvalues of the scattering matrix. Since the scattering coefficients are the sum and difference of the two eigenvalues, this approach immediately yields the entries of the scattering matrix. A simple microwave test set that allows these two eigenvalues to be measured is also described in this chapter.

1.1 *THE SCATTERING MATRIX*

The scattering matrix of a general m-port junction is defined by

$$\bar{b} = \bar{S}\bar{a} \tag{1.1}$$

where $\bar{S}$ is a square matrix that for a 2-port network has the form

$$\bar{S} = \begin{bmatrix} S_{11} & S_{12} \\ S_{21} & S_{22} \end{bmatrix} \tag{1.2}$$

The elements along the main diagonal are reflection coefficients, whereas those along the off-diagonal are transmission ones.

The vectors $\bar{a}$ and $\bar{b}$ are column matrices given by

$$\bar{a} = \begin{bmatrix} a_1 \\ a_2 \end{bmatrix} \tag{1.3}$$

$$\bar{b} = \begin{bmatrix} b_1 \\ b_2 \end{bmatrix} \tag{1.4}$$

Thus the relation between the incoming and outgoing waves for a 2-port

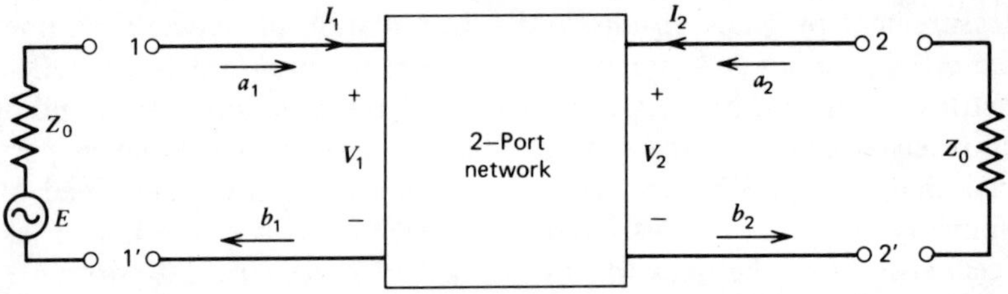

Figure 1.1. Schematic diagram indicating definition of incoming and outgoing waves for a 2-port network.

network becomes

$$b_1 = a_1 S_{11} + a_2 S_{12} \tag{1.5}$$

$$b_2 = a_1 S_{21} + a_2 S_{22} \tag{1.6}$$

This relation is given schematically in Figure 1.1.

The scattering parameters of the 2-port can be expressed in terms of the incident and reflected waves as

$$S_{11} = \left.\frac{b_1}{a_1}\right|_{a_2=0} \tag{1.7}$$

$$S_{21} = \left.\frac{b_2}{a_1}\right|_{a_2=0} \tag{1.8}$$

$$S_{12} = \left.\frac{b_1}{a_2}\right|_{a_1=0} \tag{1.9}$$

$$S_{22} = \left.\frac{b_2}{a_2}\right|_{a_1=0} \tag{1.10}$$

Figure 1.2*a* and *b* illustrates one way that the scattering parameters may be obtained experimentally.

It is assumed that a_i and b_i are normalized so that $\frac{1}{2}a_i a_i^*$ is the available power at port i and $\frac{1}{2}b_i b_i^*$ is the emergent power at the same port. For a 2-port network the a's and b's are defined by

$$a_1 = \frac{1}{2}\left(\frac{V_1}{\sqrt{R_0}} + \sqrt{R_0}\, I_1\right) \tag{1.11}$$

$$b_1 = \frac{1}{2}\left(\frac{V_1}{\sqrt{R_0}} - \sqrt{R_0}\, I_1\right) \tag{1.12}$$

$$a_2 = \frac{1}{2}\left(\frac{V_2}{\sqrt{R_0}} + \sqrt{R_0}\, I_2\right) \tag{1.13}$$

$$b_2 = \frac{1}{2}\left(\frac{V_2}{\sqrt{R_0}} - \sqrt{R_0}\, I_2\right) \tag{1.14}$$

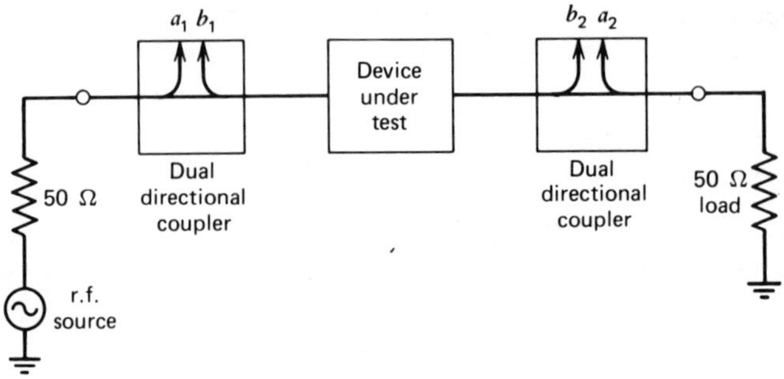

Figure 1.2a. Microwave test set for evaluating S_{11} *and* S_{21}.

To show that $\frac{1}{2}a_i a_i^*$ is the available power at port 1, it is only necessary to form the voltage V_1 in terms of the generator voltage E_1 and internal impedance R_0

$$V_1 = E_1 - R_0 I_1 \tag{1.15}$$

Substituting this value of V_1 into the definition of a_1 gives

$$a_1 = \frac{1}{2}\frac{E_1}{\sqrt{R_0}} \tag{1.16}$$

the result is

$$\tfrac{1}{2}a_1 a_1^* = \frac{E_1^2}{8R_0} \tag{1.17}$$

This is just the available power of a generator of e.m.f. E_1 and internal impedance R_0.

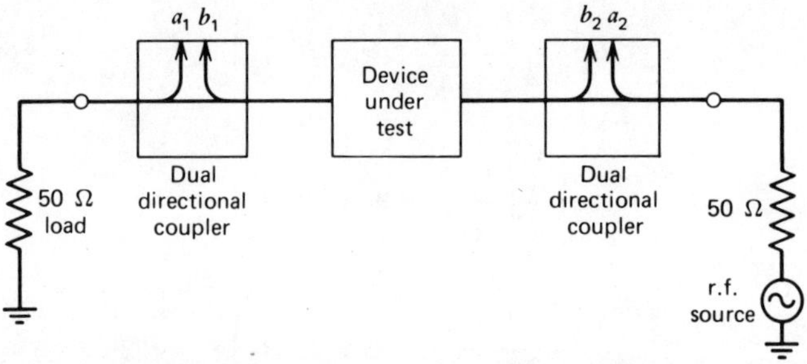

Figure 1.2b. Microwave test set for evaluating S_{22} *and* S_{12}.

To show that $\frac{1}{2} B_i b_i^*$ is the emergent power at port 2, it is necessary to combine (1.13) and (1.14) with $a_2=0$. This gives

$$b_2 = \frac{V_2}{\sqrt{R_0}} \tag{1.18}$$

Thus the power in the load is

$$\tfrac{1}{2} b_2 b_2^* = \frac{V_2^2}{2R_0} \tag{1.19}$$

The significance of the transmission parameters may now be inferred by forming S_{21} as defined by Eq. 1.8

$$S_{21} = \left.\frac{b_2}{a_1}\right|_{a_2=0} = 2\frac{V_2}{E_1} \tag{1.20}$$

so that S_{21} is the voltage transfer ratio of the network.

The meaning of the reflection parameters may also be obtained by using the definition for S_{11} given by Eq. 1.7

$$S_{11} = \left.\frac{b_1}{a_1}\right|_{a_2=0} = \frac{(V_1/\sqrt{R_0}) - I_1\sqrt{R_0}}{(V_1/\sqrt{R_0}) + I_1\sqrt{R_0}} \tag{1.21}$$

Thus

$$S_{11} = \frac{R_1 - R_0}{R_1 + R_0} \tag{1.22}$$

This is just the familiar reflection coefficient of a 1-port network. Dual relations to those above apply to S_{12} and S_{22}.

1.2 *THE SCATTERING MATRIX EIGENVALUES*

The relation between the scattering matrix and its eigenvalues can be obtained from the eigenvalue equation of the square matrix $\overline{S}$ shown schematically in Figure 1.3

$$\overline{S}\,\overline{U}_n = s_n \overline{U}_n \tag{1.23}$$

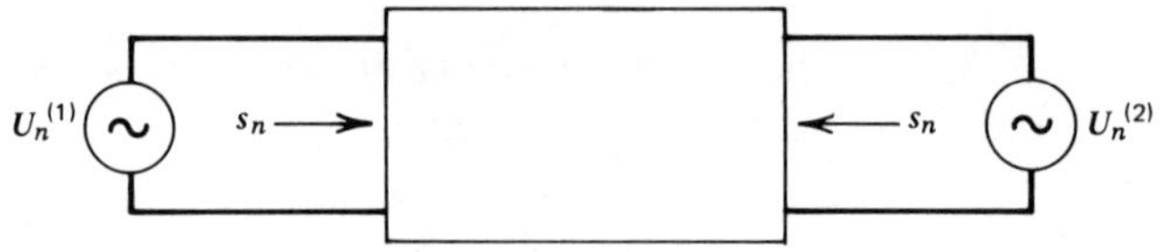

$$s_n\bar{U}_n = \bar{S}\,\bar{U}_n \qquad (\bar{b} = \bar{S}\bar{a})$$

Figure 1.3. Schematic diagram illustrating eigenvalue equation.

where $\bar{U}_n$ is an eigenvector and s_n is an eigenvalue. Through comparison with Eq. 1.1 it can be seen that $\bar{U}_n$ represents a possible excitation in the junction with the fields at the terminal planes proportional to the elements of the eigenvector, and s_n represents a reflection coefficient measured at any terminal plane. Equation 1.23 has a nonvanishing value for $\bar{U}_n$ provided

$$\det|\bar{S} - s_n\bar{I}| = 0 \qquad (1.24)$$

where $\bar{I}$ is a unit vector.

Equation 1.24 is known as the characteristic equation. The determinant given by the last equation is a polynomial of degree m. Its m roots are the m eigenvalues, of $\bar{S}$, some of which may be equal (degenerate). For a lossless junction, they lie in the complex plane with unit amplitude. These eigenvalues can be obtained once the entries of the scattering matrix are stated.

The characteristic equation for a 2-port network is

$$\begin{vmatrix} S_{11} - s_n & S_{21} \\ S_{21} & S_{11} - s_n \end{vmatrix} = 0 \qquad (1.25)$$

provided the junction is both reciprocal and symmetrical. Expanding this determinant gives

$$(S_{11} - s_n)^2 - S_{21}^2 = 0 \qquad (1.26)$$

The two roots of the characteristic equation are

$$s_1 = S_{11} + S_{21} \qquad (1.27)$$

$$s_2 = S_{11} - S_{21} \qquad (1.28)$$

Thus the eigenvalues are linear combinations of the entries of the scattering matrix.

The scattering coefficients may also be written in terms of the eigenvalues as

$$S_{11} = \frac{s_1 + s_2}{2} \tag{1.29}$$

$$S_{21} = \frac{s_1 - s_2}{2} \tag{1.30}$$

This suggests that if either set of variables is known the other may be formed. The boundary conditions of junctions may therefore be established in terms of either set of variables. If we assume that the junction is matched, the relation between the two eigenvalues can be obtained from Eq. 1.29 by

$$s_1 = -s_2 \tag{1.31}$$

which leads to

$$S_{11} = 0 \tag{1.32}$$

$$|S_{21}| = 1 \tag{1.33}$$

These two entries satisfy the unitary condition to be introduced later in this chapter.

1.3 *EIGENVECTORS*

A junction eigenvector is a unique set of incident waves determined by the symmetry of the network for which the reflection coefficient at any port is the corresponding eigenvalue of the scattering matrix. Since the eigenvectors are completely determined by the junction symmetry, a symmetrical perturbation of the junction alters the phase angles of the eigenvalues but leaves the eigenvectors unchanged. For the 2-port network illustrated in Figures 1.4*a* and 1.4*b* the two eigenvectors may be obtained by forming the eigenvalue equation given by Eq. 1.23 one at a time.

For the eigenvalue s_1, the eigenvalue equation becomes

$$\begin{bmatrix} S_{11} & S_{21} \\ S_{21} & S_{11} \end{bmatrix} \begin{bmatrix} U_1^{(1)} \\ U_1^{(2)} \end{bmatrix} = (S_{11} + S_{21}) \begin{bmatrix} U_1^{(1)} \\ U_{(1)}^{(2)} \end{bmatrix} \tag{1.34}$$

Expanding this equation gives

$$S_{11} U_1^{(1)} + S_{21} U_1^{(2)} = (S_{11} + S_{21}) U_1^{(1)} \tag{1.35}$$

$$S_{21} U_1^{(1)} + S_{11} U_1^{(2)} = (S_{11} + S_{21}) U_1^{(2)} \tag{1.36}$$

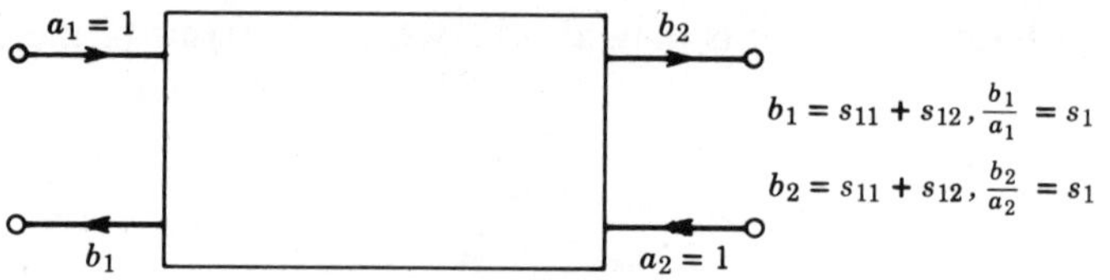

Figure 1.4a. Schematic diagram for in-phase eigensolution for 2-port network.

These two equations are satisfied provided

$$U_1^{(1)} = U_1^{(2)} = \frac{1}{\sqrt{2}} \tag{1.37}$$

This eigenvector corresponds to equal amplitude in-phase waves at ports 1 and 2 of the network in the manner illustrated in Figure 1.4*a*.

For the eigenvector s_2, the eigenvalue equation is

$$\begin{bmatrix} S_{11} & S_{21} \\ S_{21} & S_{11} \end{bmatrix} \begin{bmatrix} U_2^{(1)} \\ U_2^{(2)} \end{bmatrix} = (S_{11} - S_{21}) \begin{bmatrix} U_2^{(1)} \\ U_2^{(2)} \end{bmatrix} \tag{1.38}$$

Expanding this equation gives

$$S_{11} U_2^{(1)} + S_{21} U_2^{(2)} = (S_{11} - S_{21}) U_2^{(1)} \tag{1.39}$$

$$S_{21} U_2^{(1)} + S_{11} U_2^{(2)} = (S_{11} - S_{21}) U_2^{(2)} \tag{1.40}$$

The two equations are consistent provided

$$U_2^{(1)} = -U_2^{(2)} = \frac{1}{\sqrt{2}} \tag{1.41}$$

This solution is shown schematically in Figure 1.4*b*.

These two excitations either produce an opencircuit or a shortcircuit at the plane of symmetry of the network. The equivalent circuits or eigen-

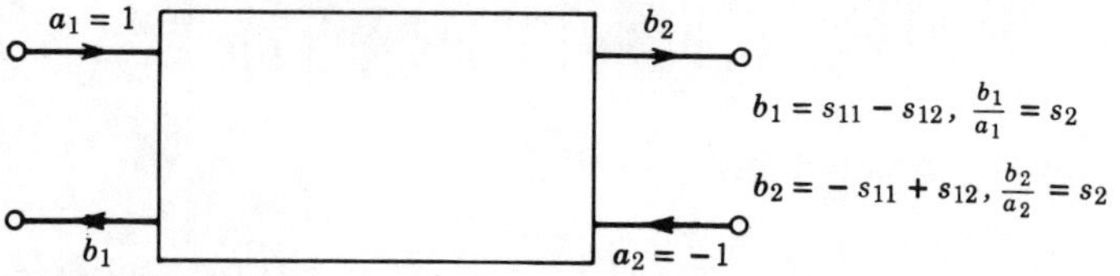

Figure 1.4b. Schematic diagram for out-of-phase eigensolution for 2-port network.

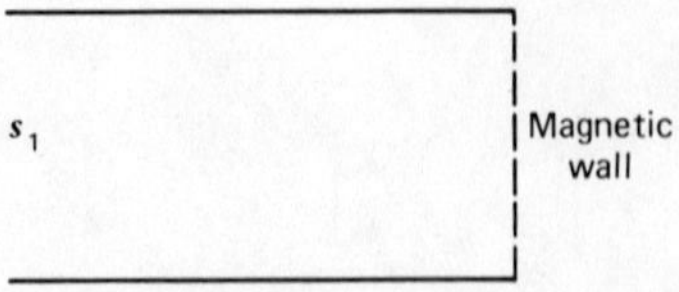

Figure 1.5a. One-port eigennetwork for in-phase eigensolution.

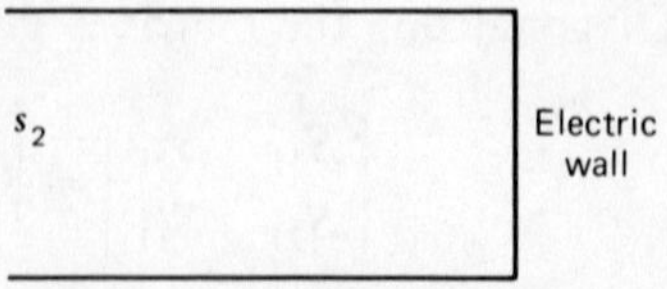

Figure 1.5b. One-port eigennetwork for out-of-phase eigensolution.

networks are therefore the 1-port opencircuited or shortcircuited transmission lines depicted in Figures 1.5*a* and 1.5*b*.

1.4 *DIAGONALIZATION OF SCATTERING MATRIX*

If the eigenvalues are known, it is possible to form the coefficients of the matrix $\bar{S}$. The relation between the two is obtained by diagonalizing $\bar{S}$. This can be done by a matrix $\bar{U}$ having for its columns the eigenvectors of $\bar{S}$

$$\bar{S}=\bar{U}\bar{\lambda}\bar{U}^{-1} \tag{1.42}$$

where $\bar{\lambda}$ is a diagonal matrix with the eigenvalues of $\bar{S}$ along its main diagonal and $\bar{U}^{-1}$ is the inverse of $\bar{U}$. If the eigenvectors of $\bar{S}$ are those obtained earlier

$$\bar{U}^{-1}=(\bar{U}^{*})^{T} \tag{1.43}$$

where $(\bar{U}^{*})^{T}$ is the transpose of the complex conjugate of $\bar{U}$. The relation between the eigenvalues and the coefficients of the scattering matrix is obtained by multiplying out Eq. 1.42.

The diagonalization procedure will now be developed for a 2-port junction. This gives the relation between the eigenvalues and the scattering coefficients of the scattering matrix. The matrix $\bar{U}$ that has the eigenvectors of $\bar{S}$ as its columns is

$$\bar{U}=\frac{1}{\sqrt{2}}\begin{bmatrix} 1 & 1 \\ 1 & -1 \end{bmatrix} \tag{1.44}$$

The diagonal matrix $\bar{\lambda}$ is

$$\bar{\lambda}=\begin{bmatrix} s_1 & 0 \\ 0 & s_2 \end{bmatrix} \tag{1.45}$$

Diagonalizing the matrix $\bar{S}$ gives

$$\begin{bmatrix} S_{11} & S_{21} \\ S_{21} & S_{11} \end{bmatrix} = \frac{1}{2}\begin{bmatrix} 1 & 1 \\ 1 & -1 \end{bmatrix}\begin{bmatrix} s_1 & 0 \\ 0 & s_2 \end{bmatrix}\begin{bmatrix} 1 & 1 \\ 1 & -1 \end{bmatrix}$$

Thus

$$S_{11} = \frac{s_1 + s_2}{2} \tag{1.46}$$

$$S_{21} = \frac{s_1 - s_2}{2} \tag{1.47}$$

which is the result obtained earlier.

1.5 *SCATTERING PARAMETERS OF 2-PORT NETWORKS*

The scattering parameters of symmetrical networks may be readily obtained from their equivalent circuits by forming their eigennetworks. These are obtained by bisecting the network and opencircuiting and shortcircuiting the exposed terminals, which is described in Chapter 2. This approach will be illustrated now in the case of a uniform section of transmission line and also for shunt and series immittances loading a transmission line.

For a uniform transmission line of electrical length θ in Figure 1.6*a*, the two eigennetworks are opencircuited and shortcircuited lines of electrical lengths $\theta/2$. The two reflection eigenvalues are therefore

$$s_1 = \frac{Z_{o/c} - Z_0}{Z_{o/c} + Z_0} \tag{1.48}$$

$$s_2 = \frac{Z_{s/c} - Z_0}{Z_{s/c} + Z_0} \tag{1.49}$$

It is observed that s_1 is associated with the opencircuited eigennetwork,

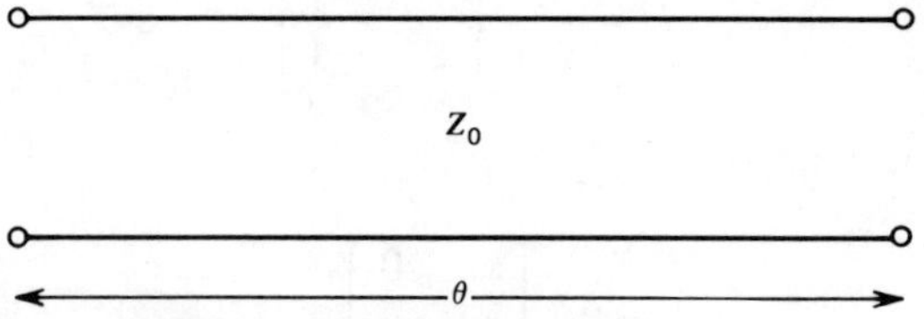

Figure 1.6a. Section of uniform line.

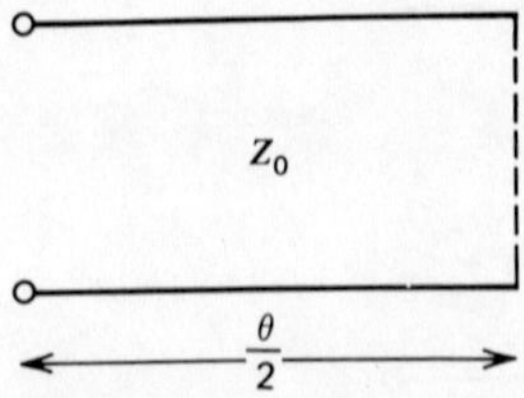

Figure 1.6b. In-phase eigennetwork for uniform section of line.

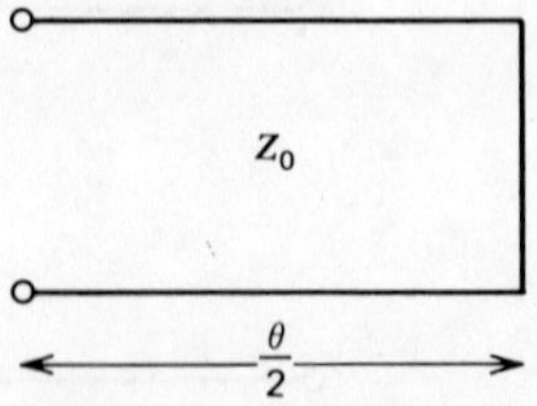

Figure 1.6c. Out-of-phase eigennetwork for uniform section of line.

since its eigenvector corresponds to in-phase waves at the two ports of the network, whereas s_2 corresponds to the shortcircuited eigennetwork, since it is related to the eigenvector having out-of-phase waves at the ports. The opencircuited and shortcircuited impedances for the eigennetworks in Figures 1.6*b* and 1.6*c* are

$$Z_{o/c} = Z_0 \coth \frac{\theta}{2} \tag{1.50}$$

$$Z_{s/c} = Z_0 \tanh \frac{\theta}{2} \tag{1.51}$$

and Z_0 is the characteristic impedance of the line.

Substituting Eqs. 1.50 and 1.51 into Eqs. 1.48 and 1.49 gives

$$s_1 = \exp(-\theta) \tag{1.52}$$

$$s_2 = -\exp(-\theta) \tag{1.53}$$

Thus

$$S_{11} = S_{22} = \frac{s_1 + s_2}{2} = 0 \tag{1.54}$$

$$S_{21} = S_{12} = \frac{s_1 - s_2}{2} = \exp(-\theta) \tag{1.55}$$

The scattering matrix for the shunt admittance loading a transmission line of characteristic admittance Y_0 in Figure 1.7*a* proceeds once more by constructing the two eigennetworks. The eigennetworks shown in Figures 1.7*b* and 1.7*c* are obtained by bisecting the network and opencircuiting and shortcircuiting the exposed terminals. The reflection eigenvalues for

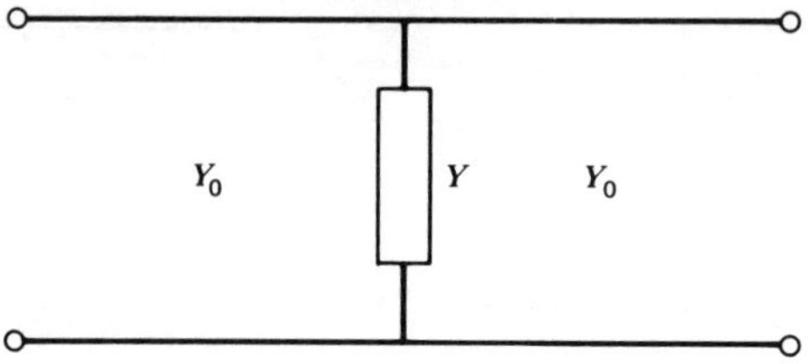

Figure 1.7a. Section of uniform line loaded with shunt network.

these two networks are

$$s_1 = \frac{Y_0 - Y_{o/c}}{Y_0 + Y_{o/c}} \tag{1.56}$$

$$s_2 = \frac{Y_0 - Y_{s/c}}{Y_0 + Y_{s/c}} \tag{1.57}$$

where

$$Y_{o/c} = \frac{Y}{2} \tag{1.58}$$

$$Y_{s/c} = \infty \tag{1.59}$$

and Y_0 is the characteristic admittance of the transmission line. Combining Eqs. 1.56 to 1.59 yields

$$S_{11} = S_{22} = \frac{s_1 + s_2}{2} = \frac{-Y}{Y + 2Y_0} \tag{1.60}$$

$$S_{21} = S_{12} = \frac{s_1 - s_2}{2} = \frac{2Y_0}{Y + 2Y_0} \tag{1.61}$$

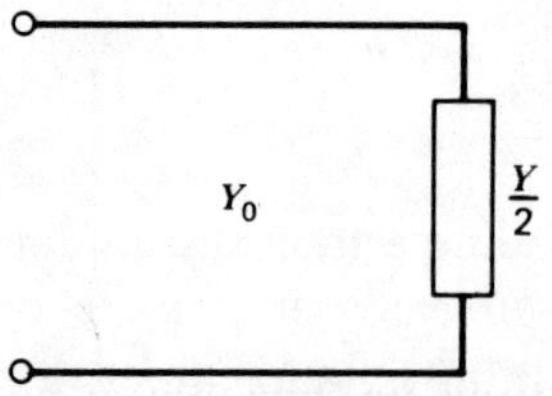

Figure 1.7b. In-phase eigennetwork for uniform line loaded with shunt network.

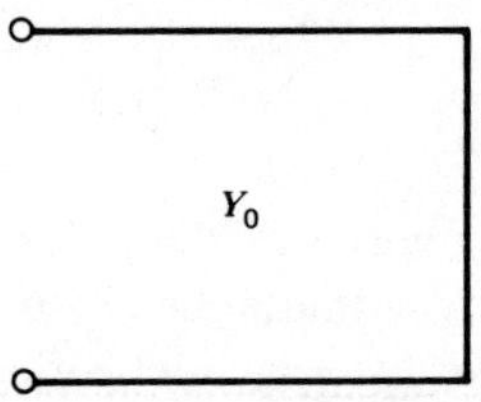

Figure 1.7c. Out-of-phase eigennetwork for uniform line loaded with shunt network.

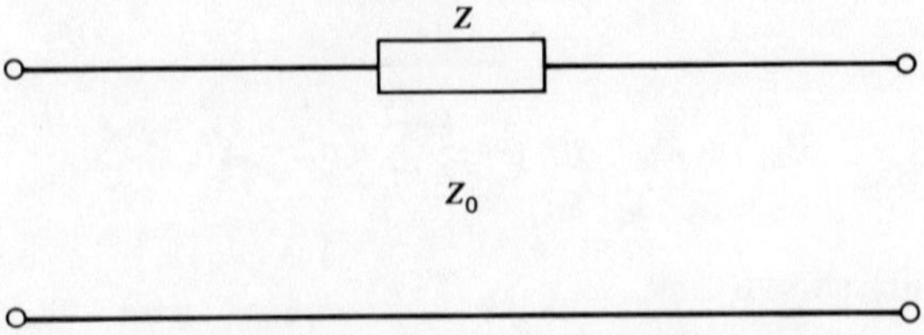

Figure 1.8a. *Section of uniform line loaded with series network.*

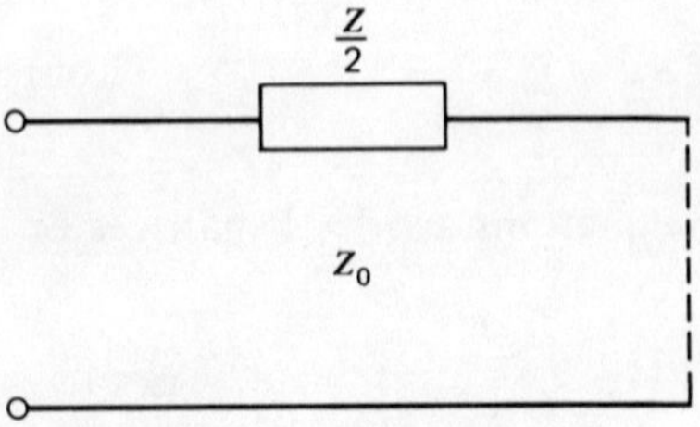

Figure 1.8b. *In-phase eigennetwork for uniform line loaded with series network.*

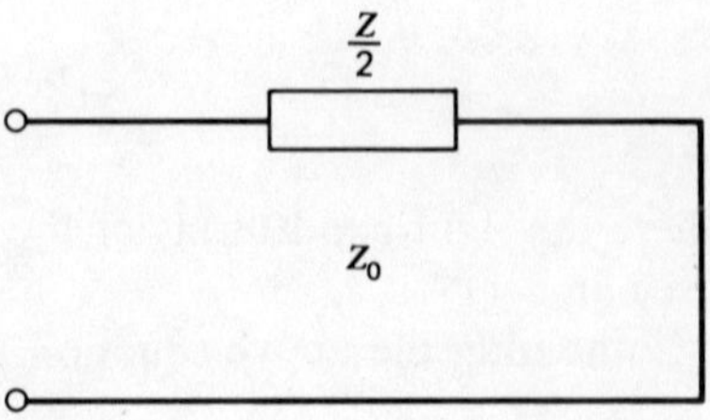

Figure 1.8c. *Out-of-phase eigennetwork for uniform line loaded with series network.*

which is the required result.

Bisecting the symmetrical network formed by the series impedance in Figure 1.8*a* leads to the two eigennetworks in Figures 1.8*b* and 1.8*c*. Following the procedure adopted for the first two situations yields the following coefficients for the scattering matrix.

$$S_{11}=S_{22}=\frac{Z}{Z+2Z_0} \tag{1.62}$$

$$S_{21}=S_{12}=\frac{2Z_0}{Z+2Z_0} \tag{1.63}$$

The derivation of the scattering matrices for T and π networks is left as an exercise for the reader.

1.6 *THE UNITARY CONDITION*

The principle of the conservation of energy states that the total power incident at all the ports of a passive junction must equal the power absorbed in the circuit plus the power emerging from the junction at all ports. This principle implies that the power dissipated P_{diss} in the network equals the difference between the input power P_{in} and the output power

P_{out},

$$P_{diss} = P_{in} - P_{out} = \frac{1}{2} \sum_i a_i a_i^* - b_i b_i^* \tag{1.64}$$

It may be readily shown that

$$\sum_i a_i^* a_i = \tilde{a}^* \bar{a} \tag{1.65}$$

$$\sum_i b_i^* b_i = \tilde{a}^* \tilde{S}^* \bar{S} \bar{a} \tag{1.66}$$

where the right-hand terms of the last two equations are 1×1 matrices or just numbers.

Combining the above equation gives

$$P_{diss} = \tfrac{1}{2} \tilde{a}^* (\bar{I} - \tilde{S}^* \bar{S}) \bar{a} \tag{1.67}$$

where $\bar{I}$ is the unit matrix, $\tilde{S}$ is the transpose of $\bar{S}$ obtained by interchanging the columns and rows of $\bar{S}$, and $\bar{S}^*$ is the complex conjugate of $\bar{S}$.

Since P_{diss} is always positive for a passive junction, the condition for a lossless junction becomes

$$\bar{I} - \tilde{S}^* \bar{S} = 0 \tag{1.68}$$

This last statement implies that the scattering matrix of a dissipationless junction is unitary. The unitary condition is widely used to establish the permissible relations between the entries of the matrix $\bar{S}$. The use of this property is illustrated in the discussion of 3-port junctions later in this text. Its significance may also be examined by applying it to form the relation between the scattering parameters of a 2-port network as will now be done. Applying the unitary condition to a 2-port network gives

$$|S_{11}|^2 + |S_{21}|^2 = 1 \tag{1.69}$$

$$|S_{22}|^2 + |S_{12}|^2 = 1 \tag{1.70}$$

$$S_{11}^* S_{12} + S_{21}^* S_{22} = 0 \tag{1.71}$$

$$S_{11} S_{12}^* + S_{21} S_{22}^* = 0 \tag{1.72}$$

The first two equations are expressions of the conservation of energy when there is an input wave at either port 1 or 2. The third or fourth equations represent the situation when both ports are excited by their eigenvectors.

This result also indicates that for a lossless network the two eigenvalues s_1 and s_2 are phasors with unit amplitude, which may be located on a unit circle, as may be verified by direct substitution into Eqs. 1.69 through 1.72

$$s_1 = \exp(-j\theta_1) \tag{1.73}$$

$$s_2 = \exp(-j\theta_2) \tag{1.74}$$

Equations 1.69 through 1.72 indicate that it is not possible to have a one-way transmission line since putting $S_{21}=0$ in these equations gives $S_{12}=0$ also.

1.7 *THE DISSIPATION MATRIX*

For a lossy network it is convenient to define a dissipation matrix to represent the dissipation within the circuit

$$\bar{Q} = \bar{I} - \tilde{S}^* \bar{S} \tag{1.75}$$

For a 2-port symmetrical junction, $\bar{Q}$ is given by

$$\bar{Q} = \begin{bmatrix} Q_{11} & -Q_{21} \\ -Q_{21}^* & Q_{11} \end{bmatrix} \tag{1.76}$$

The necessary and sufficient condition for the dissipation matrix to be positive real is that its principle minors be nonnegative

$$Q_{11} \geqslant 0 \tag{1.77}$$

$$\begin{vmatrix} Q_{11} & -Q_{21} \\ -Q_{21}^* & Q_{11} \end{vmatrix} \geqslant 0 \tag{1.78}$$

A more simple definition to this last one is also obtained by requiring that the dissipation matrix eigenvalues lie between zero and unity.

If the scattering and dissipation matrices have common eigenvectors, their eigenvalues may be related by the following theorem: If

$$\bar{Q}\bar{U}_n = q_n \bar{U}_n \tag{1.79}$$

then

$$f(\bar{Q})\bar{U}_n = f(q_n)\bar{U}_n \tag{1.80}$$

where $\overline{U}_n$ is an eigenvector and q_n is an eigenvalue. Using Eq. 1.80 in conjunction with Eq. 1.75 gives

$$q_1 = 1 - s_1 s_1^* \tag{1.81}$$

$$q_2 = 1 - s_2 s_2^* \tag{1.82}$$

The above relations show that the eigenvalues of the dissipation matrix represent the dissipation associated with each possible way of exciting the junction. These eigenvalues are real quantities that become zero when those of the scattering matrix become unity.

The dissipation matrix may be diagonalized as described for the scattering. The result is

$$Q_{11} = \frac{q_1 + q_2}{2} \tag{1.83}$$

$$Q_{21} = Q_{21}^* = \frac{q_1 - q_2}{2} \tag{1.84}$$

Combining Eqs. 1.67 and 1.75 gives

$$P_{\text{diss}} = \tfrac{1}{2}\tilde{a}^* \overline{Q} \bar{a} \tag{1.85}$$

If the junction is excited only at one input port the loss is uniquely determined by Q_{11}. If more than one port is excited at the same time, the loss of the network is also dependent on Q_{21} and the amplitude and phase of the incident waves.

For a 2-port symmetrical network

$$P_{\text{diss}} = \frac{1}{2}\begin{bmatrix} a_1^* a_2^* \end{bmatrix} \begin{bmatrix} Q_{11} & -Q_{21} \\ -Q_{21} & Q_{11} \end{bmatrix} \begin{bmatrix} a_1 \\ a_2 \end{bmatrix} \tag{1.86}$$

Thus

$$P_{\text{diss}} = \tfrac{1}{2}a_1 a_1^* Q_{11} + \tfrac{1}{2}a_2 a_2^* Q_{11} - \tfrac{1}{2}a_1 a_2^* Q_{21} - \tfrac{1}{2}a_1^* a_2 Q_{21} \tag{1.87}$$

Putting $a_1 = a_0$ and $a_2 = 0$ gives

$$P_{\text{diss}} = \tfrac{1}{2}a_0^2 Q_{11} \tag{1.88}$$

Thus, when $a_2 = 0$, Q_{11} represents the usual dissipation of the circuit.

Putting $a_1 = a_0$ and $a_2 = \mp a_0$ gives

$$P_{\text{diss}} = a_0^2(Q_{11} - Q_{21}) \tag{1.89}$$

$$P_{\text{diss}} = a_0^2(Q_{11} + Q_{21}) \tag{1.90}$$

If we rewrite the above in terms of q_1 and q_2, the dissipation in each eigennetwork becomes

$$P_{\text{diss}} = q_1 \tag{1.91}$$

$$P_{\text{diss}} = q_2 \tag{1.92}$$

The two values of dissipation of the network for input waves corresponding to its two eigenvectors are just the two eigenvalues of the dissipation matrix as expected.

Taking the shunt conductance in Figure 1.7 as an example gives

$$s_1 = \frac{Y_0 - G/2}{Y_0 + G/2} \tag{1.93}$$

$$s_2 = -1 \tag{1.94}$$

Thus

$$q_1 = 1 - s_1 s_1^* = \frac{8Y_0 G}{(G + 2Y_0)^2} \tag{1.95}$$

$$q_2 = 1 - s_2 s_2^* = 0 \tag{1.96}$$

and

$$Q_{11} = \frac{q_1 + q_2}{2} = \frac{4Y_0 G}{(G + 2Y_0)^2} \tag{1.97}$$

The complete description of the network is therefore

$$S_{11} = \frac{-G}{G + 2Y_0} \tag{1.98}$$

$$S_{21} = \frac{2Y_0}{G + 2Y_0} \tag{1.99}$$

$$Q_{11} = \frac{4Y_0 G}{(G + 2Y_0)^2} \tag{1.100}$$

These three terms represent the reflected, transmitted, and dissipated waves of the network. They are related through Eq. 1.75 by

$$Q_{11}=1-|S_{11}|^2-|S_{21}|^2 \tag{1.101}$$

1.8 *SCATTERING TRANSFER PARAMETERS*

In dealing with circuits in cascade, the scattering formalism is not the best description of the network. To overcome this difficulty, scattering transfer parameters are defined. This new matrix is obtained by rearranging the scattering relations so that the input waves a_1 and b_1 are the dependent variables and the output waves a_2 and b_2 are the independent ones. In the original $\bar{S}$ matrix, the backward waves b_1 and b_2 are dependent variables and a_1 and a_2 are the independent variables. This new matrix is known as the $\bar{T}$ matrix. The standard $\bar{S}$ matrix relates a_1, a_2 to b_1, b_2 by

$$\begin{bmatrix} b_1 \\ b_2 \end{bmatrix}=\begin{bmatrix} S_{11} & S_{12} \\ S_{21} & S_{22} \end{bmatrix}\begin{bmatrix} a_1 \\ a_2 \end{bmatrix} \tag{1.102}$$

Rearranging this matrix so that a_1 and b_1 are the dependent variables gives

$$\begin{bmatrix} b_1 \\ a_1 \end{bmatrix}=\begin{bmatrix} T_{11} & T_{12} \\ T_{21} & T_{22} \end{bmatrix}\begin{bmatrix} a_2 \\ b_2 \end{bmatrix} \tag{1.103}$$

where

$$T_{11}=S_{12}-\frac{S_{11}S_{22}}{S_{21}} \tag{1.104}$$

$$T_{12}=\frac{S_{11}}{S_{21}} \tag{1.105}$$

$$T_{21}=-\frac{S_{22}}{S_{21}} \tag{1.106}$$

$$T_{22}=\frac{1}{S_{21}} \tag{1.107}$$

A transfer matrix is also sometimes defined with the input waves as the independent variables and the output ones as the dependent variables.

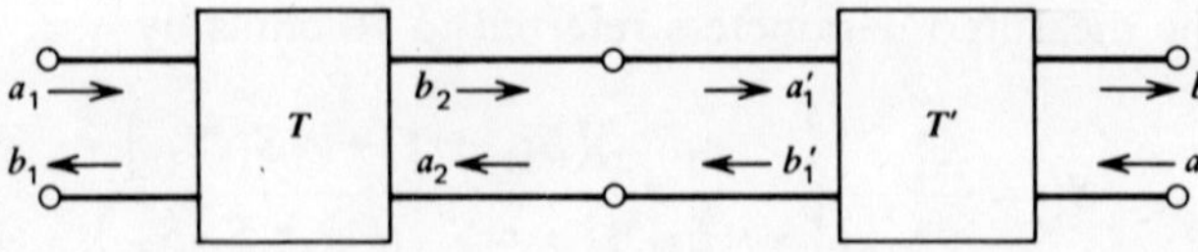

Figure 1.9. Cascade connection of two 2-port networks.

Referring to the nomenclature in Figure 1.9 the transfer matrices for the two networks are

$$\begin{bmatrix} b_1 \\ a_1 \end{bmatrix} = \begin{bmatrix} T_{11} & T_{12} \\ T_{21} & T_{22} \end{bmatrix} \begin{bmatrix} a_2 \\ b_2 \end{bmatrix} \tag{1.108}$$

$$\begin{bmatrix} b_1' \\ a_1' \end{bmatrix} = \begin{bmatrix} T_{11}' & T_{12}' \\ T_{21}' & T_{22}' \end{bmatrix} \begin{bmatrix} a_2' \\ b_2' \end{bmatrix} \tag{1.109}$$

Using

$$\begin{bmatrix} b_1' \\ a_1' \end{bmatrix} = \begin{bmatrix} a_2 \\ b_2 \end{bmatrix} \tag{1.110}$$

gives

$$\begin{bmatrix} b_1 \\ a_1 \end{bmatrix} = \begin{bmatrix} T_{11} & T_{12} \\ T_{21} & T_{22} \end{bmatrix} \begin{bmatrix} T_{11}' & T_{12}' \\ T_{21}' & T_{22}' \end{bmatrix} \begin{bmatrix} a_2' \\ b_2' \end{bmatrix} \tag{1.111}$$

Taking the ratio of b_1/a_1 gives S_{11} for the overall network.

Since matrix multiplication is not commutative, these $\bar{T}$ matrices must be multiplied in the proper order. Using the alternate definition for $\bar{T}$ parameters mentioned earlier, this matrix multiplication must be done in the reverse order.

1.9 *GENERALIZED 2-PORT SCATTERING PARAMETERS*

While the scattering parameters of networks are usually referred to 50 ohms, it is sometimes useful to define them in terms of specific load and generator impedances. For a 2-port network the new scattering parameters for arbitrary load and generator impedances may be shown[1] to be given in terms of the measured parameters referred to 50 ohms by

$$S'_{11}=\frac{A_1^*}{A_1}\frac{\left[(1-r_2S_{22})(S_{11}-r_1^*)+r_2S_{12}S_{21}\right]}{\left[(1-r_1S_{11})(1-r_2S_{22})-r_1r_2S_{12}S_{21}\right]} \tag{1.112}$$

$$S'_{12}=\frac{A_2^*}{A_1}\frac{S_{12}\left[1-|r_1|^2\right]}{\left[(1-r_1S_{11})(1-r_2S_{22})-r_1r_2S_{12}S_{21}\right]} \tag{1.113}$$

$$S'_{21}=\frac{A_1^*}{A_2}\frac{S_{21}\left[1-|r_2|^2\right]}{\left[(1-r_1S_{11})(1-r_2S_{22})-r_1r_2S_{12}S_{21}\right]} \tag{1.114}$$

$$S'_{22}=\frac{A_2^*}{A_2}\frac{\left[(1-r_1S_{11})(S_{22}-r_2^*)+r_1S_{12}S_{21}\right]}{\left[(1-r_1S_{11})(1-r_2S_{22})-r_1r_2S_{12}S_{21}\right]} \tag{1.115}$$

where

$$A_i=\frac{(1-r_i^*)}{|1-r_i|}\left(1-|r_i|^2\right)^{1/2} \tag{1.116}$$

and

$$r_i=\frac{Z_i'-Z_i}{Z_i'+Z_i^*} \qquad i=1,2 \tag{1.117}$$

1.10 *EXPERIMENTAL DETERMINATION OF MICROWAVE NETWORKS*

The theoretical determination of the physical parameters of microwave networks is beyond the topic of this work. However, the experimental characterisation of such discontinuities is always desirable if only to verify theory. This section describes one simple experimental technique that consists of measuring each of the two junction eigenvalues one at a time. This technique makes use of the fact that for an input wave at either the *H* or *E*-plane ports of a magic-tee hybrid the output waves at the two symmetrical ports are either in-phase or out-of-phase waves. Such waves correspond to the two eigenvectors of a symmetrical junction, for which the reflection coefficient at each port of the junction is the corresponding eigenvalue. The schematic diagrams of the two experimental arrangements are indicated in Figures 1.10*a* and 1.10*b*.

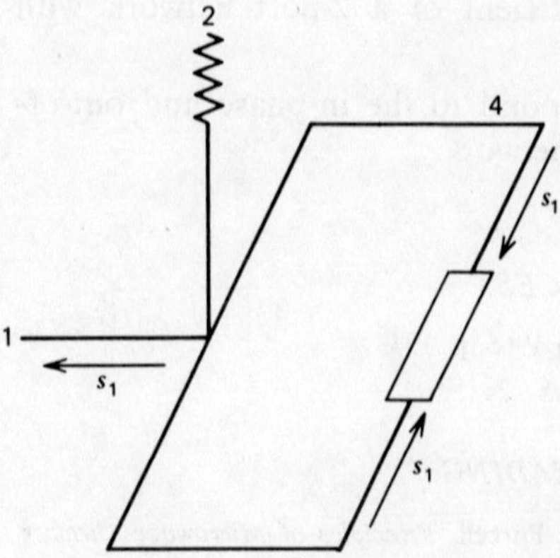

Figure 1.10a. Eigenvalue test set for evaluating s_1.

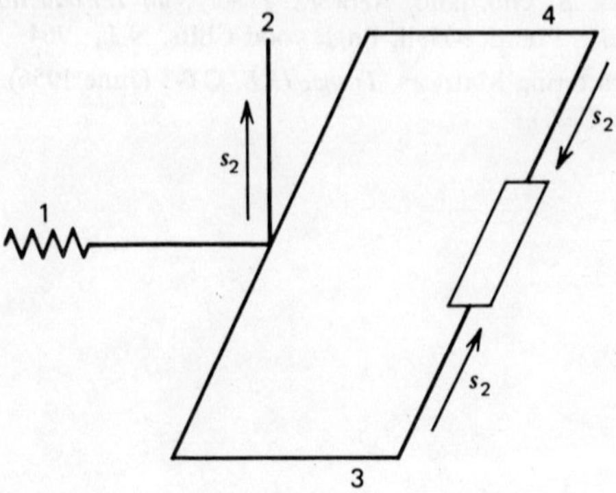

Figure 1.10b. Eigenvalue test set for evaluating s_2.

PROBLEMS

1. Obtain the scattering matrix of a series *LC* circuit.
2. Derive the scattering matrix of a shunt *LC* circuit.
3. Show that the scattering matrix of a series *LC* circuit satisfies the unitary condition.
4. Obtain the $\overline{T}$ matrices for series and shunt immittances.
5. Obtain the scattering parameters for series and shunt immittances, using the fundamental definitions given in Eqs. 1.7 to 1.10.
6. Construct the inverse of the symmetric matrix by having recourse to the theorem stated by Eqs. 1.79 and 1.80.

7. Show that the input reflection coefficient of a 2-port network with $a_2 = S_L b_2$ is $S_{11} + S_{21}^2 S_L/(1 - S_{11} S_L)$.
8. Show that Eqs. 1.71 and 1.72 correspond to the in-phase and out-of-phase inputs at ports 1 and 2 of the network.

REFERENCES

1. K. Kurokawa, *IEEE Trans.* (on MTT), March 1965, p. 194.

FURTHER READING

1. C. G. Montgomery, R. H. Dicke, and E. M. Purcell, *Principles of Microwave Circuits*, McGraw-Hill, New York, 1948.
2. J. A. Altman, *Microwave Circuits*, Van Nostrand, New York, 1964.
3. H. J. Carlin and A. B. Giordano, *Network Theory, an Introduction to Reciprocal and Nonreciprocal Circuits*, Prentice-Hall, Englewood Cliffs, N.J., 1964.
4. Special Issue on Scattering Matrices, *Trans. IRE*, **GT-3** (June 1956).

CHAPTER TWO

Immittance Matrices

Inductive and capacitive microwave reactances are widely utilized in microwave engineering for the construction of filters and for immittance matching of microwave components. Such reactances are generally obtained by introducing some physical discontinuity in the transmission line that either perturbs the electric or magnetic fields of the structure. Figure 2.1 indicates standard capacitive and inductive reactances in rectangular waveguide. In addition to the scattering description of such discontinuities, an immittance description is also useful since it leads to equivalent circuits for the junction.

The immittance matrix can be diagonalized in a similar way to the scattering matrix by noting that the eigenvectors of $\bar{S}$ are also those of $\bar{Z}$ and $\bar{Y}$, because the matrices $\bar{Z}$ and $\bar{Y}$ commute with $\bar{S}$ and the symmetry operators. Another property of the $\bar{S}$, $\bar{Z}$, and $\bar{Y}$ matrices is that their eigenvalues are related.

The procedure used in this book for relating the $\bar{S}$, $\bar{Z}$, and $\bar{Y}$ matrices consists first of finding the eigenvalues of one of them. The eigenvalues of that matrix will then be related to those of the other matrices, and these will then be used to directly construct the immittances. For example, the eigenvalues of the $\bar{S}$ matrix are related to those of the $\bar{Z}$ and $\bar{Y}$ matrices, through the 1-port relations between reflection and immittance.

A scattering matrix for a junction may sometimes have no impedance or admittance matrix (the elements become infinitely large). This situation is determined by the entries of the $\bar{S}$ matrix. If it is not possible to determine directly an immittance matrix corresponding to a given scattering matrix, a new scattering matrix can be defined by shifting the terminal planes. An immittance matrix may then be possible at the new reference plane.

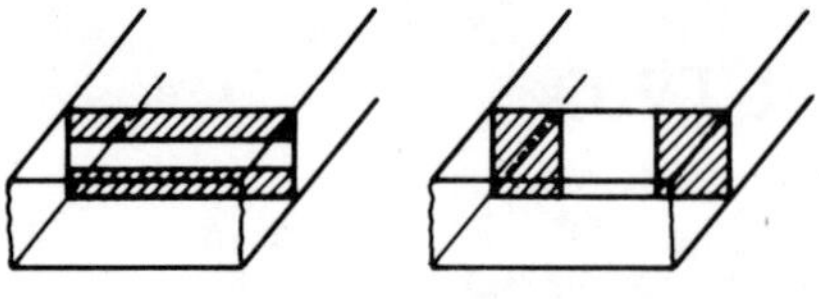

Figure 2.1. Microwave capacitive and inductive reactances.

2.1 *IMPEDANCE MATRIX*

The complete relation between voltages and currents of an m-port junction is defined through an impedance matrix given by

$$\bar{V}=\bar{Z}\bar{I} \tag{2.1}$$

where for a 2-port network the impedance matrix $\bar{Z}$ is a square matrix given by

$$\bar{Z}=\begin{bmatrix} Z_{11} & Z_{12} \\ Z_{21} & Z_{22} \end{bmatrix} \tag{2.2}$$

and the voltage and current matrices are

$$\bar{V}=\begin{bmatrix} V_1 \\ V_2 \end{bmatrix} \tag{2.3}$$

$$\bar{I}=\begin{bmatrix} I_1 \\ I_2 \end{bmatrix} \tag{2.4}$$

The entries of the $\bar{Z}$ matrix are known as opencircuit impedances defined by

$$Z_{11}=\left.\frac{V_1}{I_1}\right|_{I_2=0} \tag{2.5}$$

$$Z_{12}=\left.\frac{V_1}{I_2}\right|_{I_1=0} \tag{2.6}$$

$$Z_{22}=\left.\frac{V_2}{I_2}\right|_{I_1=0} \tag{2.7}$$

$$Z_{21}=\left.\frac{V_2}{I_1}\right|_{I_2=0} \tag{2.8}$$

For a reciprocal network

$$Z_{12} = Z_{21} \tag{2.9}$$

and for a symmetrical one

$$Z_{11} = Z_{22} \tag{2.10}$$

The eigenvalues of the $\bar{Z}$ matrix are obtained in a similar way to those of the $\bar{S}$ matrix

$$\det\left|\bar{Z} - z_n \bar{I}\right| = 0 \tag{2.11}$$

The result is

$$z_1 = Z_{11} + Z_{21} \tag{2.12}$$

$$z_2 = Z_{11} - Z_{21} \tag{2.13}$$

Rearranging the above pair of equations or diagonalizing the $\bar{Z}$ matrix gives

$$Z_{11} = \frac{z_1 + z_2}{2} \tag{2.14}$$

$$Z_{21} = \frac{z_1 - z_2}{2} \tag{2.15}$$

These eigenvalues are useful because they appear as elements in the equivalent circuit of the network and they can also readily be measured to form Z_{11} and Z_{21}.

One possible equivalent circuit for a network is the T network depicted in Figure 2.2*a*. The values of Z_1 and Z_3 of this equivalent circuit may be related to the entries of the $\bar{Z}$ matrix by applying the opencircuit definitions of the network given by Eqs. 2.5 thru 2.8

$$Z_{11} = Z_{22} = Z_1 + Z_3 \tag{2.16}$$

$$Z_{12} = Z_{21} = Z_3 \tag{2.17}$$

this leads to

$$Z_1 = Z_{11} - Z_{21} \tag{2.18}$$

$$Z_3 = Z_{21} \tag{2.19}$$

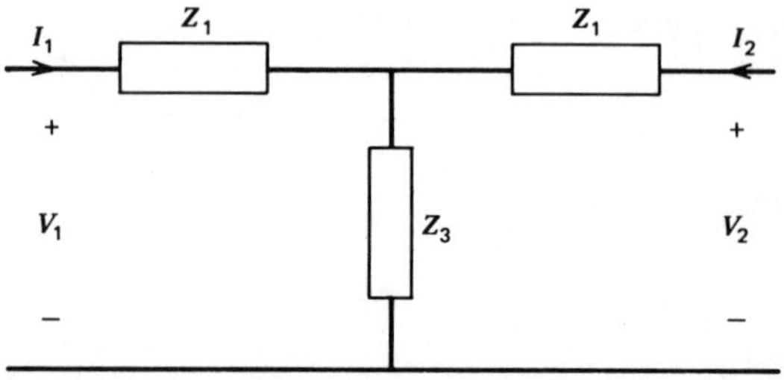

Figure 2.2a. Equivalent T *network in terms of* Z_1, Z_2, *and* Z_3.

Figure 2.2*b* depicts the equivalent circuit of the 2-port network in terms of the opencircuit parameters.

The equivalent circuit of the network can also be constructed in terms of its eigenvalues by writing Z_1 and Z_3 as

$$Z_1 = z_2 \tag{2.20}$$

$$Z_3 = \frac{z_1 - z_2}{2} \tag{2.21}$$

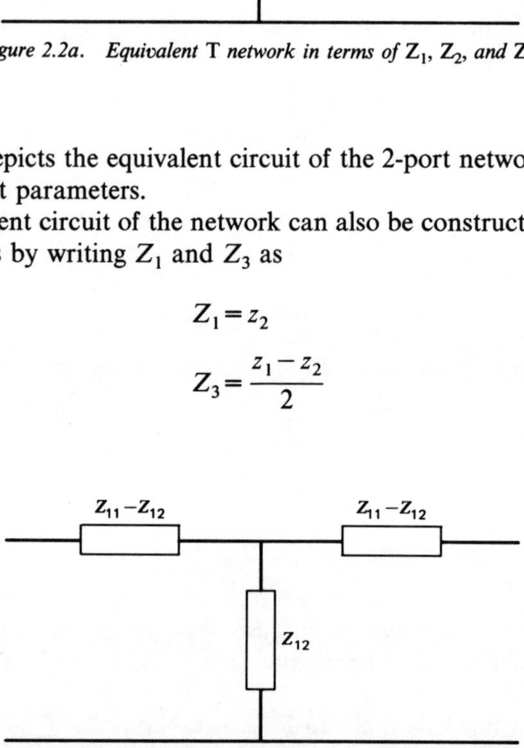

Figure 2.2b. Equivalent T network in terms of Z_{11} *and* Z_{12}.

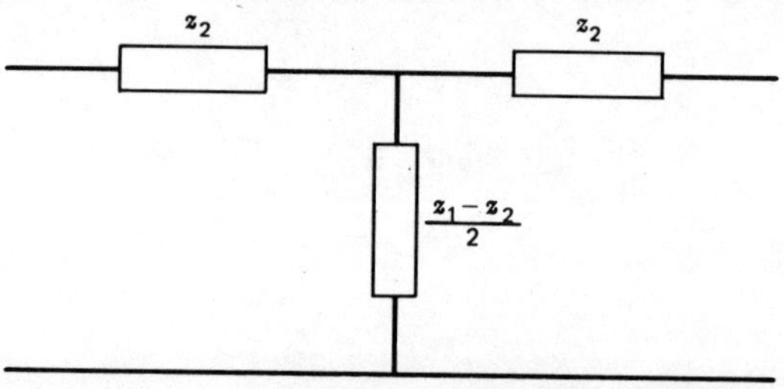

Figure 2.2c. Equivalent T *network in terms of* z_1 *and* z_2 *for symmetrical network.*

Thus the series element of the T network is just the eigenvalue z_2, whereas the shunt element is the difference between the two eigenvalues. This equivalent circuit is given in Figure 2.2*c*.

It is observed from this equivalent circuit that the input impedance at each port for in-phase voltages at the two ports is the eigenvalue z_1, whereas the input impedance at each port for out-of-phase voltages there is the eigenvalue z_2.

2.2 *THE ADMITTANCE MATRIX*

It is also possible to define an admittance matrix $\bar{Y}$ for a junction provided the inverse of the impedance matrix $\bar{Z}$ exists.

$$\bar{Y}=\bar{Z}^{-1} \tag{2.22}$$

The admittance matrix relates currents and voltages at the m ports of the junction by

$$\bar{I}=\bar{Y}\bar{V} \tag{2.23}$$

where for a 2-port network the admittance matrix $\bar{Y}$ is a square matrix given by

$$\bar{Y}=\begin{bmatrix} Y_{11} & Y_{12} \\ Y_{21} & Y_{22} \end{bmatrix} \tag{2.24}$$

and the voltage and current matrices $\bar{V}$ and $\bar{I}$ are column matrices given by Eqs. 2.3 and 2.4.

The admittance parameters are defined as

$$Y_{11}=\left.\frac{I_1}{V_1}\right|_{V_2=0} \tag{2.25}$$

$$Y_{12}=\left.\frac{I_1}{V_2}\right|_{V_1=0} \tag{2.26}$$

$$Y_{21}=\left.\frac{I_2}{V_1}\right|_{V_2=0} \tag{2.27}$$

$$Y_{22}=\left.\frac{I_2}{V_2}\right|_{V_1=0} \tag{2.28}$$

The above quantities are known as shortcircuit admittance parameters.

Symmetry and reciprocity imply

$$Y_{11}=Y_{22} \tag{2.29}$$

$$Y_{12}=Y_{21} \tag{2.30}$$

Eigenvalues can also be derived for the admittance matrix by forming the characteristic equation. The result has the symmetry of the impedance description

$$y_1=Y_{11}+Y_{21} \tag{2.31}$$

$$y_2=Y_{11}-Y_{21} \tag{2.32}$$

Diagonalizing the $\bar{Y}$ matrix as was done in the case of the $\bar{S}$ matrix or taking linear combinations of the above two equations yields

$$Y_{11}=\frac{y_1+y_2}{2} \tag{2.33}$$

$$Y_{21}=\frac{y_1-y_2}{2} \tag{2.34}$$

A π equivalent circuit is often employed to represent the shortcircuit admittance parameters. This circuit is illustrated in Figure 2.3*a*. The equivalence between the shortcircuit admittances Y_{11} and Y_{21} and the branch admittance Y_1 and Y_3 is obtained by applying the shortcircuit admittance definitions to the equivalent circuit

$$Y_{11}=Y_1+Y_3 \tag{2.35}$$

$$Y_{21}=Y_3 \tag{2.36}$$

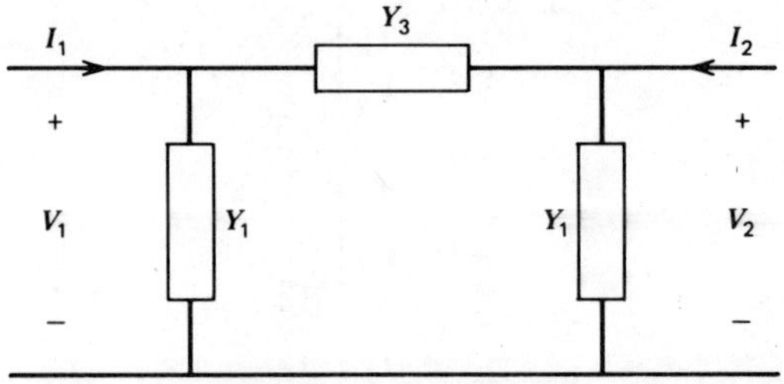

Figure 2.3a.. Equivalent π network in terms of Y_1, Y_2, and Y_3.

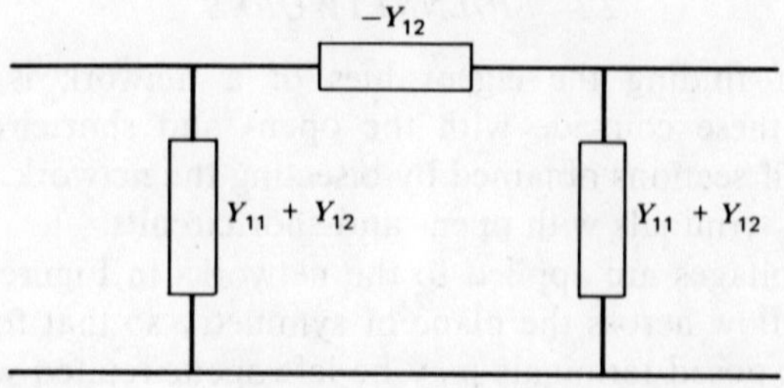

Figure 2.3b. Equivalent π network in terms of Y_{11} and Y_{12}.

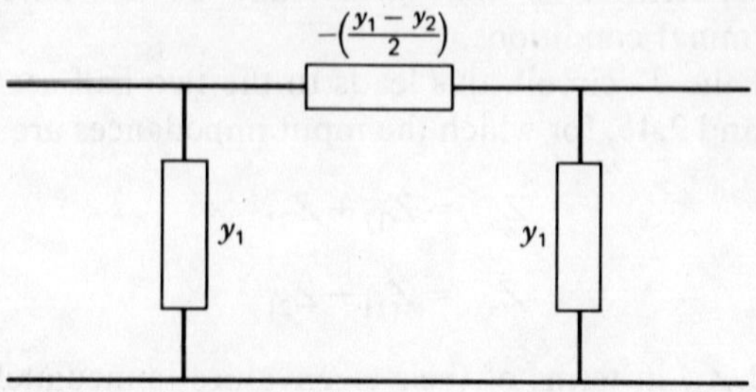

Figure 2.3c. Equivalent π network in terms of y_1 and y_2 for symmetrical network

Rearranging these two equations gives

$$Y_1 = Y_{11} - Y_{21} \tag{2.37}$$

$$Y_3 = -Y_{21} \tag{2.38}$$

The π equivalent circuit for this notation is given in Figure 2.3*b*.

This result is also given in terms of the eigenvalues of the admittance matrix as

$$Y_1 = y_2 \tag{2.39}$$

$$Y_3 = -\frac{(y_1 - y_2)}{2} \tag{2.40}$$

The equivalent circuit that applies here is given in Figure 2.3*c*. Applying the appropriate eigenvectors to this circuit gives once more the corresponding eigenvalue at each port.*

*It is observed that whereas the equivalent networks in Figures 2.2*a* and 2.3*a* give proper relations between voltage and current they may not always be realizable by LCR elements because Z_{12} and Y_{12} need not be positive real functions.

2.3 *EIGENNETWORKS*

One method of finding the eigenvalues of a network is obtained by observing that these coincide with the open- and shortcircuited immittances of the half sections obtained by bisecting the network and terminating the exposed terminals with open- and shortcircuits.

If in-phase voltages are applied to the networks in Figures 2.2 and 2.3, no current can flow across the plane of symmetry so that for this type of excitation the exposed terminals may be left opencircuited without affecting the conditions at the input terminals. Similarly, if out-of-phase voltages are applied to the networks, the terminals crossing the plane of symmetry have the same potential so that these may be shortcircuited without affecting the terminal conditions.

In the case of the *T*- circuit, this leads to the two half-sections depicted in Figures 2.4*a* and 2.4*b*, for which the input impedances are

$$Z_{o/c} = Z_{11} + Z_{21} \tag{2.41}$$

$$Z_{s/c} = Z_{11} - Z_{21} \tag{2.42}$$

Writing Z_{11} and Z_{21} in terms of their eigenvalues immediately gives

$$Z_{o/c} = z_1 \tag{2.43}$$

$$Z_{s/c} = z_2 \tag{2.44}$$

This result implies that the two half-sections are just the two eigennetworks associated with the impedance matrix.

For the π circuit the two half-sections become those in Figures 2.5*a* and 2.5*b* for which the input admittances are

$$Y_{o/c} = Y_{11} + Y_{21} \tag{2.45}$$

$$Y_{s/c} = Y_{11} - Y_{21} \tag{2.46}$$

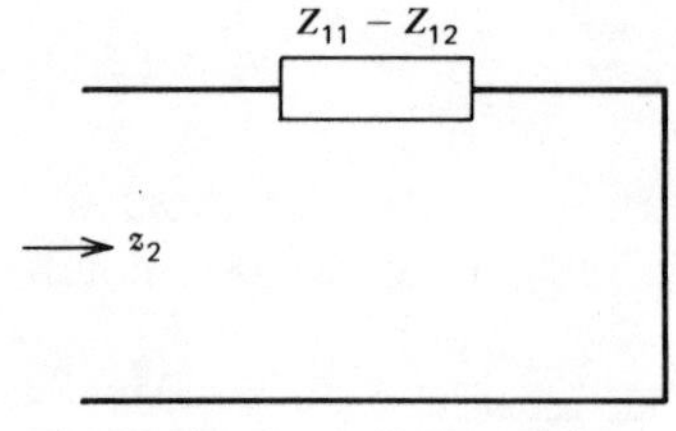

Figure 2.4b. Out-of-phase eigennetwork for T*-symmetrical network.*

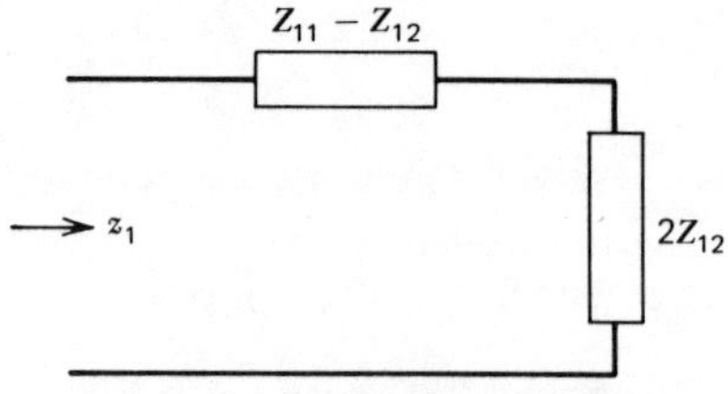

Figure 2.4a. In-phase eigennetwork for T*-symmetrical network.*

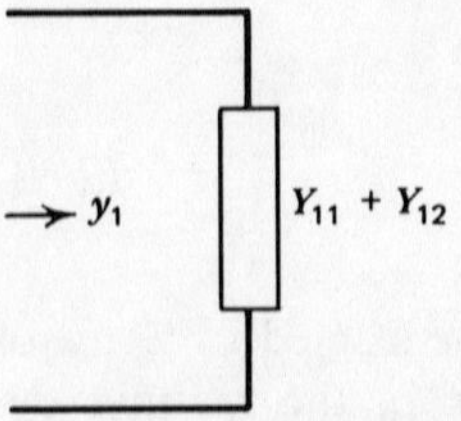

Figure 2.5a. In-phase eigennetwork for π-symmetrical network.

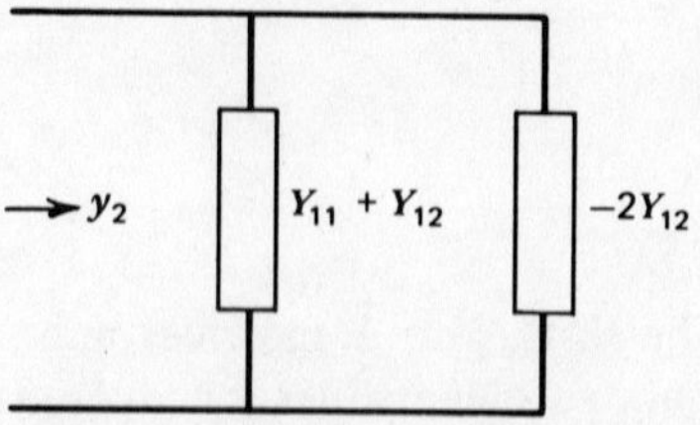

Figure 2.5b. Out-of-phase eigennetwork for π-symmetrical network.

In terms of the admittance eigenvalues, these become

$$Y_{o/c} = y_1 \tag{2.47}$$

$$Y_{s/c} = y_2 \tag{2.48}$$

Thus the open- and shortcircuited admittances of the half sections are again just the two eigenvalues of the network.

2.4 *EIGENVALUES OF SCATTERING AND IMMITTANCE MATRICES*

The eigenvalues of the $\bar{S}$, $\bar{Z}$, or $\bar{Y}$ matrices may be found in the usual way by forming the eigenvalue equation. However, if any one set of them is known, the others may be obtained from it in a rather simple way. This may be done by observing that the eigenvalues are the usual reflection, impedance, and admittance associated with 1-port networks for which the standard 1-port relations apply. The relations between the eigenvalues, therefore, take the following standard forms

$$s_n = \frac{Z_n - Z_0}{Z_n + Z_0} \tag{2.49}$$

$$\frac{Z_n}{Z_0} = \frac{1 + s_n}{1 - s_n} \tag{2.50}$$

$$s_n = \frac{Y_0 - Y_n}{Y_0 + Y_n} \tag{2.51}$$

$$\frac{Y_n}{Y_0} = \frac{1 - s_n}{1 + s_n} \tag{2.52}$$

and

$$Z_n = \frac{1}{Y_n} \tag{2.53}$$

The $\bar{S}$, $\bar{Z}$, and $\bar{Y}$ matrices may now be directly constructed in terms of the above eigenvalues by diagonalizing the matrices in the manner described in Chapter 1. This immediately leads to the matrices defined by Eqs. 1.29 and 1.30 and Eqs. 2.14, 2.15, 2.33, and 2.34.

2.5 *RELATION BETWEEN $\bar{S}$, $\bar{Z}$, AND $\bar{Y}$ MATRICES*

It has been observed that networks may be described either by their scattering or immittance matrices. Other descriptions are also possible but are not dealt with here. In some situations it is desirable to relate one description to another. Therefore, the relation between them will now be obtained. One method already discussed consists of diagonalizing the matrices by using the relation between their eigenvalues. However, two other approaches are also useful. The first relies on a theorem that relates the form of the matrix relation to that of the corresponding eigenvalue one, whereas the other starts with the initial definition of the scattering variables in terms of voltage and current. The correspondence between the scattering and impedance matrices will first be derived by starting with the definition of the normalized scattering column matrices in terms of the normalized voltage and current matrices

$$\bar{a} = \tfrac{1}{2}(\bar{V} + \bar{i}) \tag{2.54}$$

$$\bar{b} = \tfrac{1}{2}(\bar{V} - \bar{i}) \tag{2.55}$$

which have the form of Eqs. 1.11 through 1.14.

Rewriting the above equations gives

$$\bar{V} = \bar{a} + \bar{b} \tag{2.56}$$

$$\bar{i} = \bar{a} - \bar{b} \tag{2.57}$$

Substituting the relation between $\bar{a}$ and $\bar{b}$ into the last two equations gives

$$\bar{V} = (\bar{I} + \bar{S})\bar{a} \tag{2.58}$$

$$\bar{i} = (\bar{I} - \bar{S})\bar{a} \tag{2.59}$$

Replacing $\bar{V}$ by $\bar{Z}\bar{i}$

$$\bar{Z}\bar{i}=(\bar{I}+\bar{S})\bar{a} \tag{2.60}$$

$$\bar{i}=(\bar{I}-\bar{S})\bar{a} \tag{2.61}$$

Hence

$$(\bar{I}+\bar{S})=\bar{Z}(\bar{I}-\bar{S}) \tag{2.62}$$

Postmultiplying both sides by $(\bar{I}-\bar{S})^{-1}$ gives

$$\bar{Z}=(\bar{I}+\bar{S})(\bar{I}-\bar{S})^{-1} \tag{2.63}$$

The $\bar{Z}$ matrix will exist provided the inverse matrix $(\bar{I}-\bar{S})^{-1}$ exists. This condition is satisfied if the matrix $(\bar{I}-\bar{S})$ is nonsingular. This requires that the determinant of the matrix should not be zero.

$$|\bar{I}-\bar{S}|\neq 0 \tag{2.64}$$

The above condition is determined by the entries of the scattering matrix.

A more convenient method for obtaining this last result is to make use of the following theorem, which applies provided the matrices have common eigenvectors. If

$$\bar{S}\bar{U}_n=s_n\bar{U}_n \tag{2.65}$$

then

$$f(\bar{S})\bar{U}_n=f(s_n)\bar{U}_n \tag{2.66}$$

This theorem will first be applied to obtain the relation between the $\bar{S}$ and $\bar{Z}$ matrices.

The relation between s_n and z_n is given by Eq. 2.49 as

$$s_n=\frac{z_n-1}{z_n+1} \tag{2.67}$$

$\bar{S}$ is therefore related to $\bar{Z}$ in the same manner.

$$\bar{S}=\frac{(\bar{Z}-\bar{I})}{(\bar{Z}+\bar{I})} \tag{2.68}$$

Applying this theorem to the other 1-port relationships gives the following matrix results

$$\bar{Z}=\frac{\bar{I}+\bar{S}}{\bar{I}-\bar{S}} \tag{2.69}$$

$$\bar{S}=\frac{\bar{I}-\bar{Y}}{\bar{I}+\bar{Y}} \tag{2.70}$$

$$\bar{Y}=\frac{\bar{I}-\bar{S}}{\bar{I}+\bar{S}} \tag{2.71}$$

and

$$\bar{Z}=\bar{Y}^{-1} \tag{2.72}$$

Dissipation may also be accounted for in these equations by using the dissipation matrix.

The relation between the $\bar{S}$, $\bar{Z}$, and $\bar{Y}$ elements are listed in Table 2.1.

Table 2.1

S parameters in terms of Y and Z parameters	*Y and Z parameters in terms of S parameters*
$S_{11}=\dfrac{(Z_{11}-1)(Z_{22}+1)-Z_{12}Z_{21}}{(Z_{11}+1)(Z_{22}+1)-Z_{12}Z_{21}}$	$Z_{11}=\dfrac{(1+S_{11})(1-S_{22})+S_{12}S_{21}}{(1-S_{11})(1-S_{22})-S_{12}S_{21}}$
$S_{12}=\dfrac{2Z_{12}}{(Z_{11}+1)(Z_{22}+1)-Z_{12}Z_{21}}$	$Z_{12}=\dfrac{2S_{12}}{(1-S_{11})(1-S_{22})-S_{12}S_{21}}$
$S_{21}=\dfrac{2Z_{21}}{(Z_{11}+1)(Z_{22}+1)-Z_{12}Z_{21}}$	$Z_{21}=\dfrac{2S_{21}}{(1-S_{11})(1-S_{22})-S_{12}S_{21}}$
$S_{22}=\dfrac{(Z_{11}+1)(Z_{22}-1)-Z_{12}Z_{21}}{(Z_{11}+1)(Z_{22}+1)-Z_{12}Z_{21}}$	$Z_{22}=\dfrac{(1+S_{22})(1-S_{11})+S_{12}S_{21}}{(1-S_{11})(1-S_{22})-S_{12}S_{21}}$
$S_{11}=\dfrac{(1-Y_{11})(1+Y_{22})+Y_{22}Y_{21}}{(1+Y_{11})(1+Y_{22})-Y_{12}Y_{21}}$	$Y_{11}=\dfrac{(1+S_{22})(1-S_{11})+S_{12}S_{21}}{(1+S_{11})(1+S_{22})-S_{12}S_{21}}$
$S_{12}=\dfrac{-2Y_{12}}{(1+Y_{11})(1+Y_{22})-Y_{22}Y_{21}}$	$Y_{12}=\dfrac{-2S_{12}}{(1+S_{11})(1+S_{22})-S_{12}S_{21}}$
$S_{21}=\dfrac{-2Y_{21}}{(1+Y_{11})(1+Y_{22})-Y_{12}Y_{21}}$	$Y_{21}=\dfrac{-2S_{21}}{(1+S_{11})(1+S_{22})-S_{12}S_{21}}$
$S_{22}=\dfrac{(1+Y_{11})(1-Y_{22})+Y_{12}Y_{21}}{(1+Y_{11})(1+Y_{22})-Y_{12}Y_{21}}$	$Y_{22}=\dfrac{(1+S_{11})(1-S_{22})+S_{12}S_{21}}{(1+S_{22})(1+S_{11})-S_{12}S_{21}}$

2.6 *EQUIVALENT CIRCUIT OF UNIFORM TRANSMISSION LINE*

The theory developed so far will now be applied to construct the equivalent circuit of a uniform section of transmission line one meter long having a propagation constant γ and characteristic impedance Z_0. The schematic of the transmission line is illustrated in Figure 2.6*a*, whereas the two eigennetworks obtained by bisecting the network are depicted in Figures 2.6*b* and 2.6*c*. The impedance eigenvalues of the two eigennetworks are

$$z_1 = Z_{o/c} = Z_0 \coth\frac{\gamma l}{2} \tag{2.73}$$

$$z_2 = Z_{s/c} = Z_0 \tanh\frac{\gamma l}{2} \tag{2.74}$$

The impedance matrix is now defined by

$$Z_{11} = \frac{z_1 + z_2}{2} = \frac{Z_0}{\tanh \gamma l} \tag{2.75}$$

$$Z_{21} = \frac{z_1 - z_2}{2} = \frac{Z_0}{\sinh \gamma l} \tag{2.76}$$

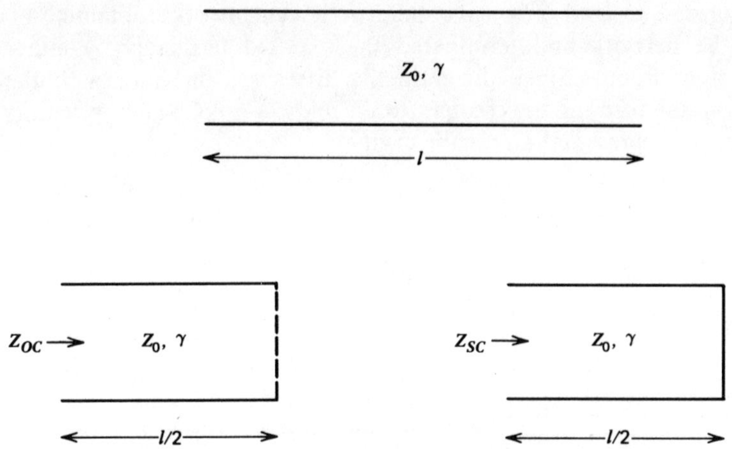

Figure 2.6. Schematic diagram for uniform transmission line.

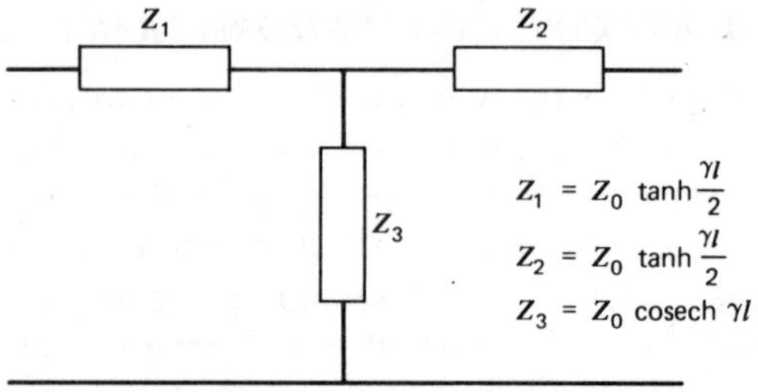

Figure 2.7. Equivalent circuit for uniform transmission line.

and the equivalent circuit is given by

$$Z_1 = Z_2 = Z_{11} - Z_{21} = Z_0 \tanh \frac{\gamma l}{2} \tag{2.77}$$

$$Z_3 = Z_{21} = \frac{Z_0}{\sinh \gamma l} \tag{2.78}$$

This equivalent circuit is shown in Figure 2.7.

2.7 *EQUIVALENT CIRCUIT OF TRANSMISSION CAVITY RESONATOR*

The equivalent circuit of the cavity resonator described in Chapter 6 can be readily obtained by the method outlined in this section. This resonator consists of a half-wave transmission line terminated at both ends by inductive irises. Figure 2.8*a* shows the schematic diagram of this resonator, while Figures 2.8*b* and 2.8*c* give the two eigennetworks obtained by bisecting the network and terminated the exposed terminals by either open- or shortcircuits. Since the inductive irises are in shunt with the transmission line forming the resonant waveguide, a π equivalent circuit is the appropriate representation for the cavity.

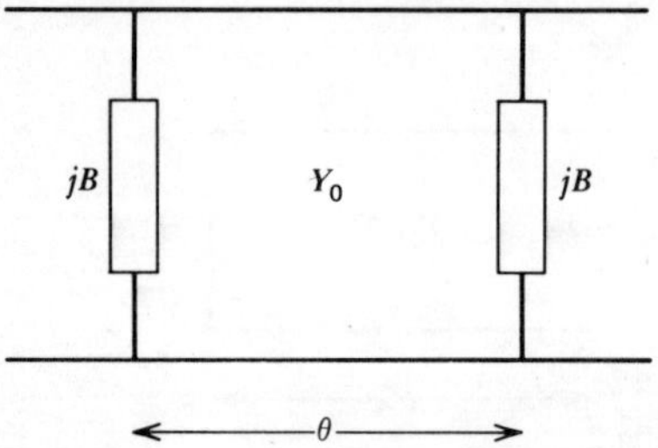

Figure 2.8. Schematic diagram for cavity resonator.

Figure 2.8. (continued)

In the case of the in-phase eigennetwork in Figure 2.8*b* the eigenadmittance is

$$y_1 = Y_{o/c} = jB + Y_0 \tanh\frac{\gamma l}{2} \tag{2.79}$$

whereas for the out-of-phase eigennetwork in Figure 2.8*c* it is

$$y_2 = Y_{s/c} = jB + Y_0 \coth\frac{\gamma l}{2} \tag{2.80}$$

Thus

$$Y_{11} = Y_{22} = \frac{y_1 + y_2}{2} = jB + \frac{Y_0}{\tanh \gamma l} \tag{2.81}$$

$$Y_{12} = Y_{21} = \frac{y_1 - y_2}{2} = \frac{Y_0}{\sinh \gamma l} \tag{2.82}$$

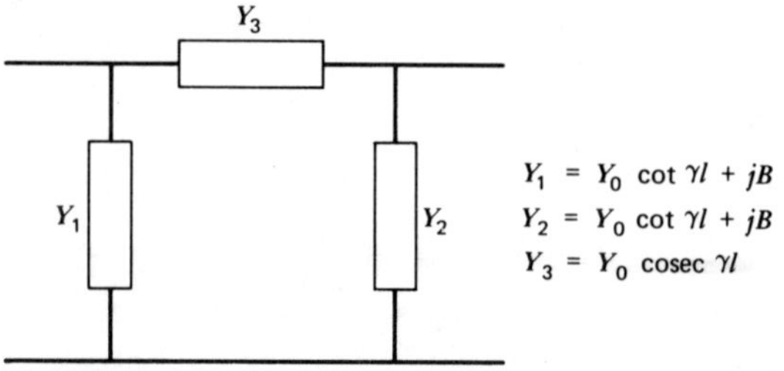

Figure 2.9. Equivalent circuit of transmission cavity resonator.

where Y_0 and γ are the characteristic admittance and complex propagation constant of the waveguide used to construct the resonator.

Figure 2.9 gives the equivalent π circuit for this device.

PROBLEMS

1. Demonstrate that there is no $\bar{Z}$ matrix for a series network and no $\bar{Y}$ matrix for a shunt one.
2. Show that the relations between the $\bar{S}$, $\bar{Z}$, and $\bar{Y}$ matrices indicated in Table 2.1 satisfy the eigenvalue relations for symmetrical matrices.
3. Derive the *ABCD* transfer matrix between V_1, i_1, and V_2, i_2 from the opencircuited impedance matrix by first reversing the polarity of i_2.
4. Show that $Z_{in} = Z_{11} - Z_{12}^2/(Z_{11} + Z_L)$.
5. Construct the π equivalent circuit of a uniform transmission line.

FURTHER READING

1. C. G. Montgomery, R. J. Dicke, and E. M. Purcell, *Principles of Microwave Circuits*, McGraw Hill, New York, 1948.
2. Special Issue on Scattering Matrices, *Trans. IRE*, GT,3 (June 1956).

CHAPTER THREE

Directional Couplers

The basic directional coupler consists of two waveguides coupled together by means of small slots in such a manner that an incident wave on one waveguide is partly transferred to the other with particular directional properties. One such arrangement is depicted in Figure 3.1. It is defined as a 4-port device having an input port, two mutually isolated output ports, and one port isolated from the input port. It has also the property that all of its ports are matched. Such networks are used as elements in power monitors and reflectometers. A special class of directional couplers of particular importance are the so called 3-dB hybrids in which the input wave is equally divided between the two output ports with either a 90- or 180-deg phase difference. Hybrid networks are widely used as elements in balanced mixers, microwave discriminators and switches, to name but a few applications.

The two most important parameters that describe the performance of the directional coupler are its coupling and directivity. Referring to Figure 3.1 the coupling is defined as the ratio of the primary input power to the secondary output power

$$\text{coupling (dB)} = 10\log_{10}\left(\frac{P_1}{P_4}\right)$$

Directivity is the ratio of total forward coupled power at port 4 to the total backscattered power at port 2

$$\text{Directivity(dB)} = 10\log_{10}\left(\frac{P_4}{P_2}\right)$$

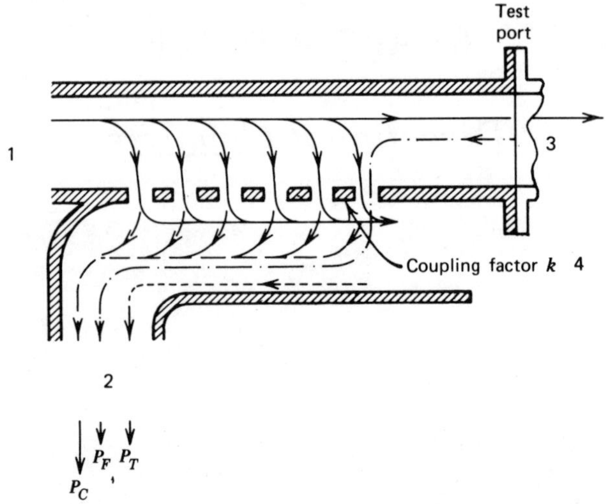

Figure 3.1. Schematic diagram of directional coupler.

Ideally, the backscattered power is zero, but in practice it consists of contributions from the discontinuities of the coupling holes (P_C), from the discontinuity of the termination at port 4 (P_T), and from the discontinuity of the transition or flange at port 3 (P_F). The backscatter from the coupling holes may be minimized by spacing them a quarter wave apart. Since in a reflectometer arrangement the backscattered signal and the reflected one from the component under test at port 3 add vectorially, the directivity of the coupler gives a measure of the smallest discontinuity that can be measured. The directivity is usually of the order of 45–55 dB.

The properties of directional couplers are almost exclusively expressed in terms of the scattering matrix, and this is the approach adopted in this text.

3.1 *SCATTERING MATRIX OF DIRECTIONAL COUPLER*

The directional coupler is a 4-port device having an input port, two mutually isolated output ports, and one port isolated from the input one. Also, all its ports are matched. The relation between the entries of the scattering matrix for such a junction is obtained by using the unitary condition subject to the above boundary conditions and symmetry and reciprocity conditions.

The general scattering matrix for the 4-port junction is

$$\bar{S}=\begin{bmatrix} S_{11} & S_{12} & S_{13} & S_{14} \\ S_{21} & S_{22} & S_{23} & S_{24} \\ S_{31} & S_{32} & S_{33} & S_{34} \\ S_{41} & S_{42} & S_{43} & S_{44} \end{bmatrix} \tag{3.1}$$

where the port designation is illustrated in Figure 3.1. Using the fact that one port is isolated from the input port gives

$$S_{12}=S_{21}=S_{34}=S_{43}=0 \tag{3.2}$$

and the scattering matrix becomes

$$\bar{S}=\begin{bmatrix} S_{11} & 0 & S_{13} & S_{14} \\ 0 & S_{22} & S_{23} & S_{24} \\ S_{31} & S_{32} & S_{33} & 0 \\ S_{41} & S_{42} & 0 & S_{44} \end{bmatrix} \tag{3.3}$$

Assuming that all ports are matched gives

$$S_{11}=S_{22}=S_{33}=S_{44} \tag{3.4}$$

which yields

$$\bar{S}=\begin{bmatrix} 0 & 0 & S_{13} & S_{14} \\ 0 & 0 & S_{23} & S_{24} \\ S_{31} & S_{32} & 0 & 0 \\ S_{41} & S_{42} & 0 & 0 \end{bmatrix} \tag{3.5}$$

Making use of the fact that the directional coupler is reciprocal requires that the scattering matrix be symmetric about the main diagonal.

$$\begin{aligned} S_{13}&=S_{31} \\ S_{14}&=S_{41} \\ S_{23}&=S_{32} \\ S_{24}&=S_{42} \end{aligned} \tag{3.6}$$

which leads to the following scattering matrix for the directional coupler involving only four entries

$$\bar{S}=\begin{bmatrix} 0 & 0 & S_{13} & S_{14} \\ 0 & 0 & S_{23} & S_{24} \\ S_{13} & S_{23} & 0 & 0 \\ S_{14} & S_{24} & 0 & 0 \end{bmatrix} \tag{3.7}$$

The symmetry of the junction may now be used to further reduce the number of entries in the last matrix. If the junction is completely symmetric, the following relations apply

$$\begin{aligned} S_{13} &= S_{24} \\ S_{14} &= S_{23} \end{aligned} \tag{3.8}$$

The final matrix therefore involves only two independent variables.

$$\bar{S}=\begin{bmatrix} 0 & 0 & S_{13} & S_{14} \\ 0 & 0 & S_{14} & S_{13} \\ S_{13} & S_{14} & 0 & 0 \\ S_{14} & S_{13} & 0 & 0 \end{bmatrix} \tag{3.9}$$

To establish whether such a junction is realizable, it is necessary to apply the unitary condition discussed in Chapter 1 to the above matrix.

$$\bar{S}\,(\bar{S}^*)^T=\bar{I} \tag{3.10}$$

which implies that the junction is lossless.

In terms of the original variables, the result is

$$\begin{bmatrix} 0 & 0 & S_{13} & S_{14} \\ 0 & 0 & S_{14} & S_{13} \\ S_{13} & S_{14} & 0 & 0 \\ S_{14} & S_{13} & 0 & 0 \end{bmatrix}\begin{bmatrix} 0 & 0 & S_{13}^* & S_{14}^* \\ 0 & 0 & S_{14}^* & S_{13}^* \\ S_{13}^* & S_{14}^* & 0 & 0 \\ S_{14}^* & S_{13}^* & 0 & 0 \end{bmatrix}=\begin{bmatrix} 1 & 0 & 0 & 0 \\ 0 & 1 & 0 & 0 \\ 0 & 0 & 1 & 0 \\ 0 & 0 & 0 & 1 \end{bmatrix} \tag{3.11}$$

This gives

$$|S_{13}|^2+|S_{14}|^2=1 \tag{3.12}$$

$$S_{13}S_{14}^*+S_{13}^*S_{14}=0 \tag{3.13}$$

The first equation satisfies energy conservation, whereas the second suggests that one possible solution at a suitable pair of terminals is

$$S_{13}=\alpha \tag{3.14}$$

$$S_{14}=j\beta \tag{3.15}$$

where α and β are real numbers.

The matrix of the symmetrical directional coupler is therefore

$$\bar{S}=\begin{bmatrix} 0 & 0 & \alpha & j\beta \\ 0 & 0 & j\beta & \alpha \\ \alpha & j\beta & 0 & 0 \\ j\beta & \alpha & 0 & 0 \end{bmatrix} \tag{3.16}$$

The above development indicates that all directional couplers are perfectly matched. One important property of the symmetrical junction is that there is a 90-deg phase difference between the waves in the two output ports.

3.2 *HYBRID JUNCTIONS*

A special class of directional couplers are the 3-dB hybrid ones, for which there is equal power division between ports 3 and 4. The symmetric 3-dB coupler is in fact a hybrid junction. Two possible forms of such hybrids are indicated in Figures 3.2*a* and 3.2*b*.

The scattering matrices for these hybrids may be obtained by starting with Eq. 3.7 and introducing the appropriate symmetries. The short slot hybrid in Figure 3.2*a* has the symmetry studied earlier so that Eq. 3.16

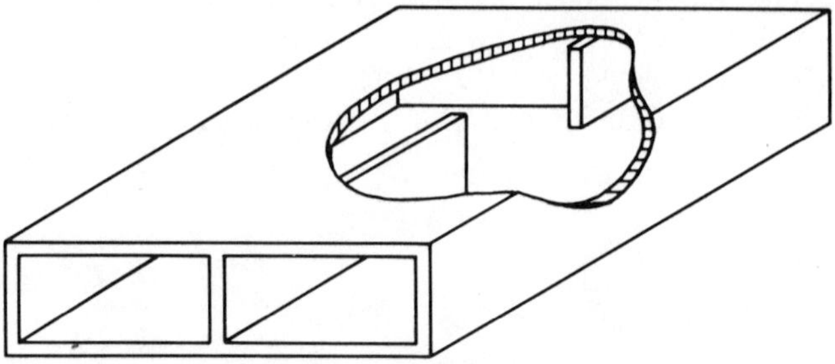

Figure 3.2a. Sidewall hybrid waveguide directional coupler.

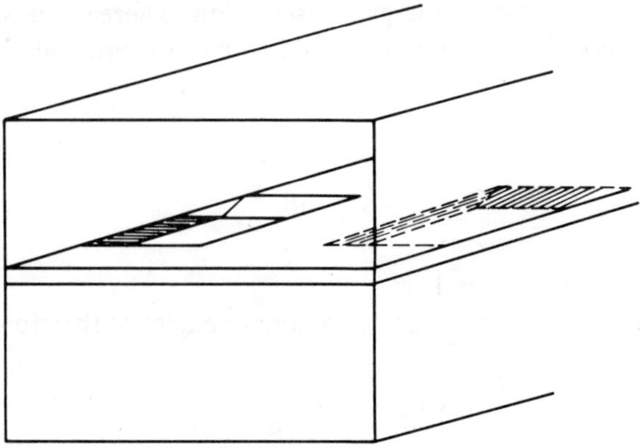

Figure 3.2b. Topwall hybrid waveguide directional coupler.

applies with $\alpha=\beta=0.707$.

$$\bar{S}=\frac{1}{\sqrt{2}}\begin{bmatrix} 0 & 0 & 1 & j \\ 0 & 0 & j & 1 \\ 1 & j & 0 & 0 \\ j & 1 & 0 & 0 \end{bmatrix} \tag{3.17}$$

Another class of hybrids with different symmetry properties are the so-called magic tee ones illustrated in Figure 3.3. The symmetry that applies to the configuration in Figure 3.3*a* may be inferred by examining the symmetries existing in the *H*-plane and *E*-plane tees in Figures 3.4*a* and 3.4*b*.

For the *H*-plane tee junction the two waves are in phase in the two waveguides so that

$$S_{13}=S_{14}=\frac{1}{\sqrt{2}} \tag{3.18}$$

applies in Eq. 3.7 provided port 1 is matched. For the *E*-plane tee junction the two waves are 180-deg out of phase in the two waveguides. Hence

$$S_{23}=-S_{24}=\frac{1}{\sqrt{2}} \tag{3.19}$$

applies in Eq. 3.7 provided port 2 is matched.

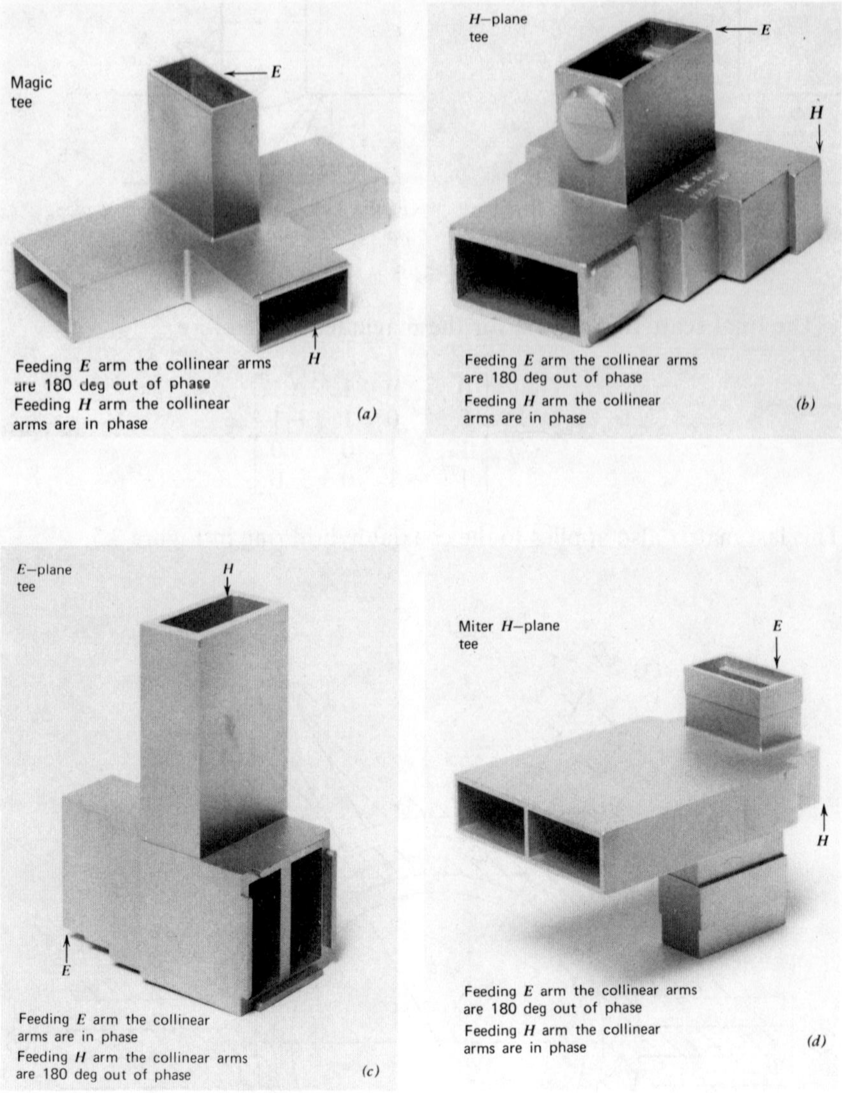

Figure 3.3. Magic-tee hybrid waveguide directional couplers. (*Courtesy M.D.L.*).

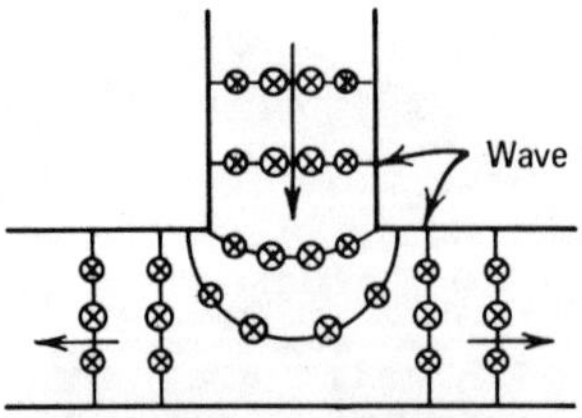

Figure 3.4a. Coupling in simple H*-plane waveguide* T *junction.*

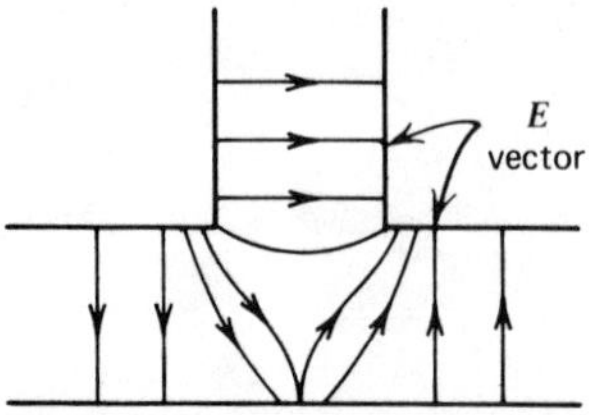

Figure 3.4b. Coupling in simple E*-plane waveguide* T *junction.*

The final scattering matrix for the magic tee is therefore

$$\bar{S} = \frac{1}{\sqrt{2}} \begin{bmatrix} 0 & 0 & 1 & 1 \\ 0 & 0 & 1 & -1 \\ 1 & 1 & 0 & 0 \\ 1 & -1 & 0 & 0 \end{bmatrix} \tag{3.20}$$

This last matrix also applies to the coaxial hybrid ring in Figure 3.5.

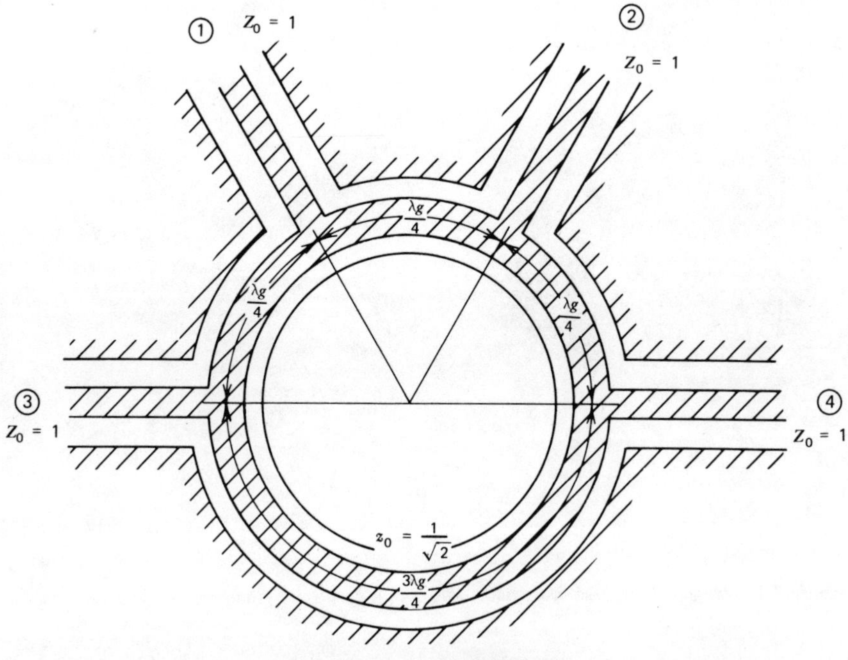

Figure 3.5. Coaxial hybrid ring.

3.3 *EVEN AND ODD MODES THEORY OF DIRECTIONAL COUPLERS*

Additional information about the behavior of junctions may normally be obtained by forming the eigenvalues of the scattering matrix. These eigenvalues are the reflection coefficients of the junction associated with the four possible ways of exciting the network that will give the eigenvalues at any port. However, in the case of the symmetrical directional coupler, a linear combination of even and odd mode transmission and reflection parameters are often employed instead. The even mode excitation is obtained by applying equal waves at ports 1 and 2 of the network, whereas the odd mode excitation is obtained by applying out-of-phase waves there. These two situations are illustrated in Figures 3.6*a* and 3.6*b*.

For these two excitations the plane of symmetry for the even mode becomes an opencircuit, whereas the plane of symmetry for the odd mode becomes a shortcircuit, and the 4-port can be analyzed from the superposition of a pair of 2-ports each operating in its respective mode.

In what follows the entries of the scattering matrix will be written in terms of the odd and even mode parameters. This may be done by taking each set of field patterns one at a time as the input waves of the directional coupler and constructing the output ones.

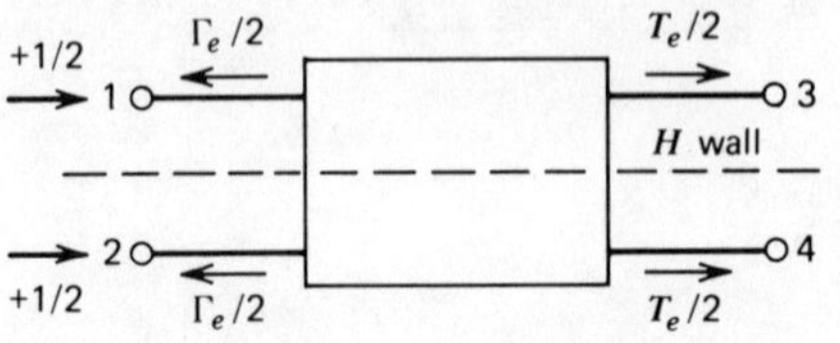

Figure 3.6a. *Schematic of directional coupler with even mode excitation.*

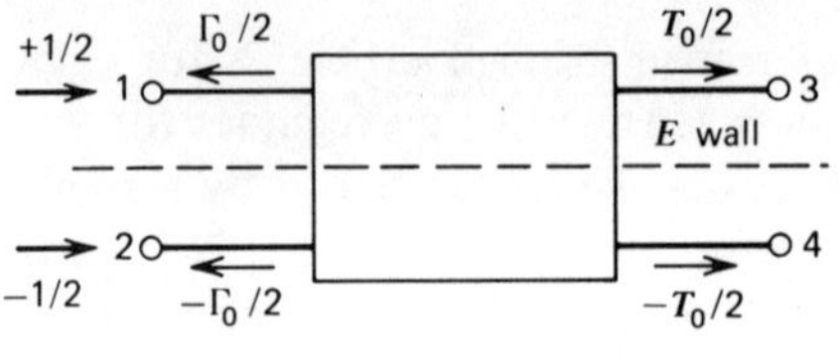

Figure 3.6b. *Schematic of directional coupler with odd mode excitation.*

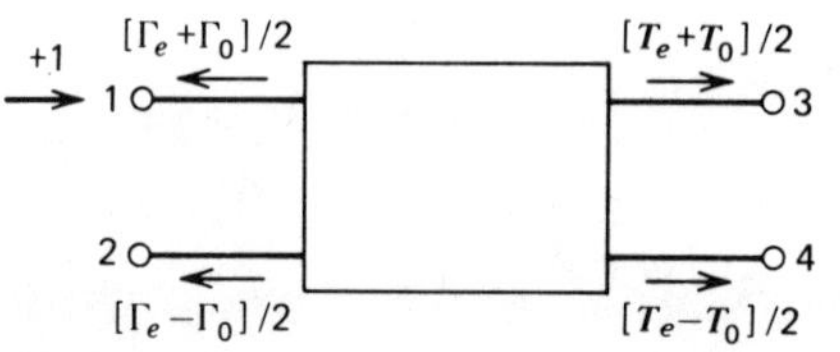

Figure 3.6c. *Schematic of directional coupler with single input.*

For the even mode excitation the input/output relation of the directional coupler is

$$\begin{bmatrix} b_1 \\ b_2 \\ b_3 \\ b_4 \end{bmatrix} = \begin{bmatrix} S_{11} & S_{12} & S_{13} & S_{14} \\ S_{12} & S_{11} & S_{14} & S_{13} \\ S_{13} & S_{14} & S_{11} & S_{12} \\ S_{14} & S_{13} & S_{12} & S_{11} \end{bmatrix} \begin{bmatrix} \frac{1}{2} \\ \frac{1}{2} \\ 0 \\ 0 \end{bmatrix} \tag{3.21}$$

The above scattering matrix assumes that the device is reciprocal and symmetrical, but no assumption is made about the boundary conditions of the ideal directional coupler. Expanding the above matrix relation gives

$$b_1 = \frac{S_{11} + S_{12}}{2} \tag{3.22}$$

$$b_2 = \frac{S_{11} + S_{12}}{2} \tag{3.23}$$

$$b_3 = \frac{S_{13} + S_{14}}{2} \tag{3.24}$$

$$b_4 = \frac{S_{13} + S_{14}}{2} \tag{3.25}$$

Even mode reflection and transmission coefficients may now be defined for each waveguide as follows:

$$\Gamma_e = \frac{b_1}{a_1} = \frac{b_2}{a_2} = S_{11} + S_{12} \tag{3.26}$$

$$\tau_e = \frac{b_3}{a_1} = \frac{b_4}{a_2} = S_{13} + S_{14} \tag{3.27}$$

where Γ_e and τ_e are the reflection and transmission coefficients for each waveguide. Since there is no coupling between the two waveguides for this set of incident waves the coupled waveguides may be replaced by a single waveguide with an even mode field pattern.

For the odd mode excitation the input/output relation of the network becomes

$$\begin{bmatrix} b_1 \\ b_2 \\ b_3 \\ b_4 \end{bmatrix} = \begin{bmatrix} S_{11} & S_{12} & S_{13} & S_{14} \\ S_{12} & S_{11} & S_{14} & S_{13} \\ S_{13} & S_{14} & S_{11} & S_{12} \\ S_{14} & S_{13} & S_{12} & S_{11} \end{bmatrix} \begin{bmatrix} \frac{1}{2} \\ -\frac{1}{2} \\ 0 \\ 0 \end{bmatrix} \tag{3.28}$$

Thus

$$b_1 = \frac{S_{11} - S_{12}}{2} \tag{3.29}$$

$$b_2 = \frac{-S_{11} + S_{12}}{2} \tag{3.30}$$

$$b_3 = \frac{S_{13} - S_{14}}{2} \tag{3.31}$$

$$b_4 = \frac{-S_{13} + S_{14}}{2} \tag{3.32}$$

Odd mode reflection and transmission coefficients for each waveguide are in this case defined by

$$\Gamma_o = \frac{b_1}{a_1} = \frac{b_2}{a_2} = S_{11} - S_{12} \tag{3.33}$$

$$\tau_o = \frac{b_3}{a_1} = \frac{b_4}{a_2} = S_{13} - S_{14} \tag{3.34}$$

The reflection and transmission coefficients are again identical for each waveguide section so that the 4-port network may be once more replaced by a 2-port one for this excitation. The above two solutions may now be combined to give

$$S_{11} = \frac{\Gamma_e + \Gamma_o}{2} \tag{3.35}$$

$$S_{12} = \frac{\Gamma_e - \Gamma_o}{2} \tag{3.36}$$

$$S_{13} = \frac{\tau_e + \tau_o}{2} \tag{3.37}$$

$$S_{14} = \frac{\tau_e - \tau_o}{2} \tag{3.38}$$

This result suggests that the boundary conditions of an ideal directional coupler may be obtained in one of two ways, as will now be seen.

In the first definition the odd and even mode reflection coefficients are zero while their transmission ones are different. These boundary conditions lead to the scattering matrix of the ideal directional coupler defined in (3.9)

$$S_{11}=0 \tag{3.39}$$

$$S_{12}=0 \tag{3.40}$$

$$S_{13}=\frac{\tau_e+\tau_o}{2} \tag{3.41}$$

$$S_{14}=\frac{\tau_e-\tau_o}{2} \tag{3.42}$$

A second solution, which also satisfies the definition of a directional coupler, is obtained provided the two reflection and transmission coefficients are $\Gamma_e=-\Gamma_o$ and $\tau_e=\tau_o$. The result is

$$S_{11}=0 \tag{3.43}$$

$$S_{12}=\Gamma_e \tag{3.44}$$

$$S_{13}=\tau_e \tag{3.45}$$

$$S_{14}=0 \tag{3.46}$$

This type of directional coupler is known as a backward wave coupler.

3.4 *OPERATION OF SIDEWALL AND TOPWALL HYBRIDS*

The use of even and odd mode parameters will now be used to illustrate the design of the sidewall and topwall hybrids in Figure 3.2. To do this it is necessary to establish the even and odd mode field patterns for each hybrid configuration and obtain their propagation constants. A possible set of field patterns for the sidewall hybrid is depicted in Figures 3.7*a* and 3.7*b*. It is seen that an electric wall may be introduced at the plane of symmetry of the coupled waveguide in the case of the even mode excitation, whereas a magnetic wall can be introduced there for the odd mode excitation. The two boundary problems to be solved are, therefore, rectangular waveguides of width a with one of their sidewalls replaced by either an electric or magnetic wall. Fortunately, there is no need to tackle these two problems since the two field patterns in this situation are TE_{10} and TE_{20} in double-width rectangular waveguide for which the transmission

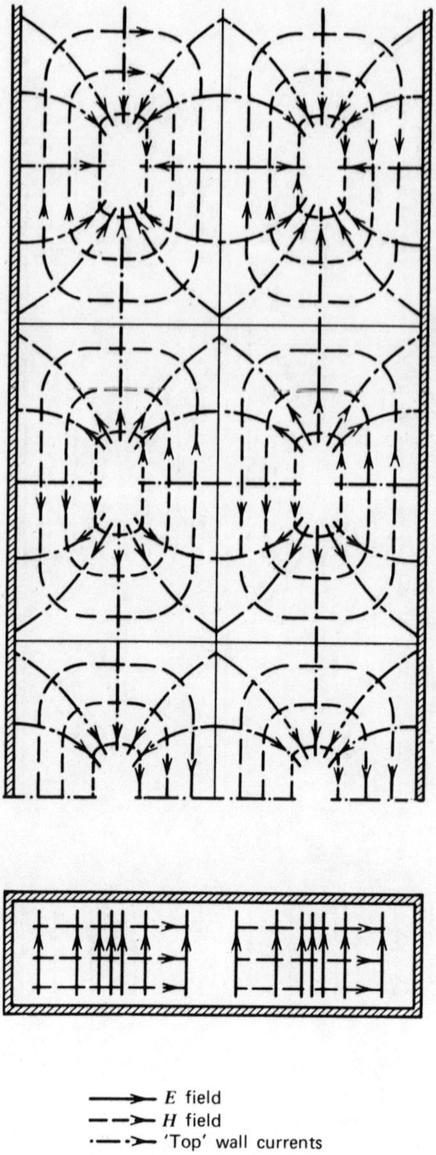

Figure 3.7a. Even mode field patterns for sidewall hybrid (electric wall).

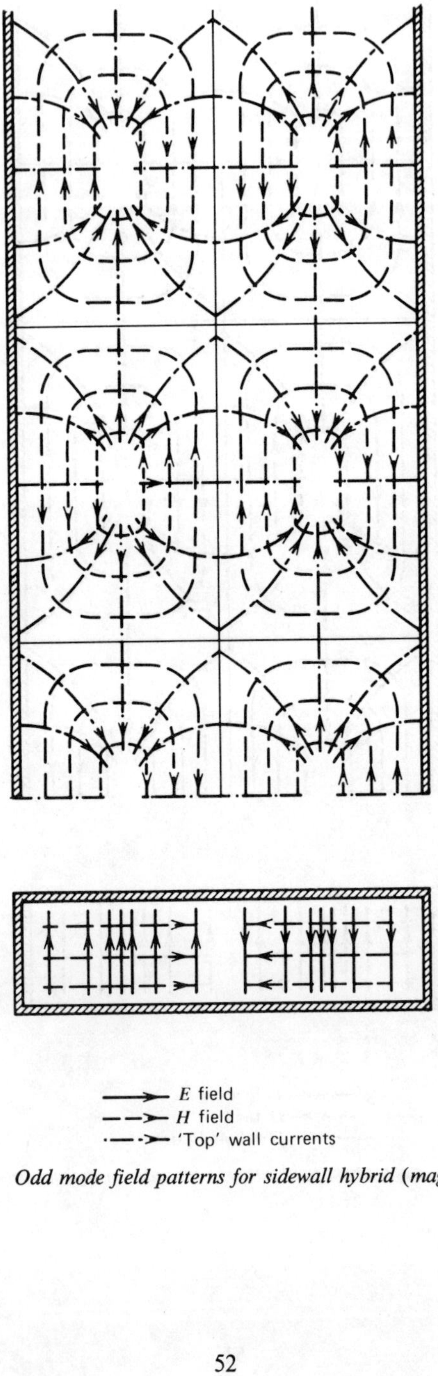

Figure 3.7b. Odd mode field patterns for sidewall hybrid (magnetic wall).

coefficients are

$$\tau_e = \exp - j\frac{2\pi L}{\lambda_{ge}} \tag{3.47}$$

$$\tau_o = \exp - j\frac{2\pi L}{\lambda_{go}} \tag{3.48}$$

$$\left(\frac{2\pi}{\lambda_{ge}}\right)^2 = \left(\frac{2\pi}{\lambda_0}\right)^2 - \left(\frac{\pi}{2a}\right)^2 \tag{3.49}$$

$$\left(\frac{2\pi}{\lambda_{go}}\right)^2 = \left(\frac{2\pi}{\lambda_0}\right)^2 - \left(\frac{\pi}{2a}\right)^2 \tag{3.50}$$

In the above equations L is the length of the coupling slot and a is the wide dimension of the rectangular waveguides. The entries for the scattering matrix of the directional coupler are now given with the help of Eqs. 3.39 through 3.42 by

$$S_{11} = 0 \tag{3.51}$$

$$S_{12} = 0 \tag{3.52}$$

$$S_{13} = \frac{\tau_e + \tau_o}{2} = \frac{\exp - j2\pi L/\lambda_{ge} + \exp - j2\pi L/\lambda_{go}}{2} \tag{3.53}$$

$$S_{14} = \frac{\tau_e - \tau_o}{2} = \frac{\exp - j2\pi L/\lambda_{ge} - \exp - j2\pi L/\lambda_{go}}{2} \tag{3.54}$$

Taking out a common factor

$$\exp\left[-\frac{1}{2}\left(\frac{2\pi}{\lambda_{ge}} + \frac{2\pi}{\lambda_{go}}\right)L\right] \tag{3.55}$$

gives

$$S_{11} = 0 \tag{3.56}$$

$$S_{12} = 0 \tag{3.57}$$

$$S_{13} = \cos\left[\left(\frac{2\pi}{\lambda_{ge}} - \frac{2\pi}{\lambda_{go}}\right)\frac{L}{2}\right]\exp\frac{-j}{2}\left(\frac{2\pi}{\lambda_{ge}} + \frac{2\pi}{\lambda_{go}}\right)L \tag{3.58}$$

$$S_{14} = -j\sin\left[\left(\frac{2\pi}{\lambda_{ge}} - \frac{2\pi}{\lambda_{go}}\right)\frac{L}{2}\right]\exp\frac{-j}{2}\left(\frac{2\pi}{\lambda_{ge}} + \frac{2\pi}{\lambda_{go}}\right)L \tag{3.59}$$

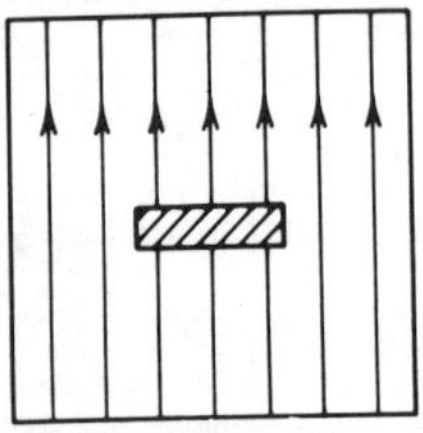

Figure 3.8a. Even mode field patterns for topwall hybrid.

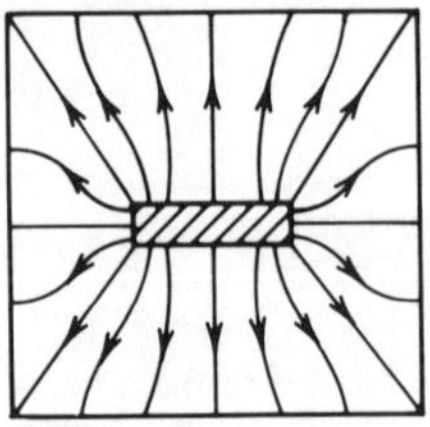

Figure 3.8b. Odd mode field patterns for topwall hybrid.

It is observed that this final result satisfies the unitary condition defined in Eqs. 3.12 through 3.13.

This result indicates that the power transfer between the two waveguides is periodic. For a 3-dB hybrid coupler $S_{13} = S_{14} = 0.707$ and the coupling length becomes

$$\left(\frac{2\pi}{\lambda_{ge}} - \frac{2\pi}{\lambda_{go}}\right)L = \frac{\pi}{2} \tag{3.60}$$

A similar development applies to the topwall hybrid except that the even and odd mode field patterns shown in Figures 3.8*a* and 3.8*b* are now those of a TEM transmission line and a TE_{10} waveguide.

3.5 *EIGENVALUE THEORY OF SIDEWALL WAVEGUIDE HYBRID*

Although the analysis of symmetrical directional couplers is usually given in terms of odd and even mode variables the eigenvalue approach may also be utilized. This will now be demonstrated for the sidewall 3-dB hybrid in Figure 3.2*a* described by Eq. 3-16. The characteristic equation for the sidewall hybrid is

$$\begin{vmatrix} -s_n & 0 & 0.707 & j0.707 \\ 0 & -s_n & j0.707 & 0.707 \\ 0.707 & j0.707 & -s_n & 0 \\ j0.707 & 0.707 & 0 & -s_n \end{vmatrix} = 0 \tag{3.61}$$

Expanding this equation gives

$$s_n^4 + 1 = 0 \tag{3.62}$$

The roots of the characteristic equation are

$$s_1 = \exp +j45 \tag{3.63}$$

$$s_2 = \exp -j45 \tag{3.64}$$

$$s_3 = -\exp +j45 \tag{3.65}$$

$$s_4 = -\exp -j45 \tag{3.66}$$

Forming the eigenvalue equation $\overline{S}\overline{U}_n = s_n \overline{U}_n$ for each of these eigenvalues gives the eigenvectors as

$$\overline{U}_1 = \frac{1}{\sqrt{4}} \begin{bmatrix} 1 \\ 1 \\ 1 \\ 1 \\ 1 \end{bmatrix} \tag{3.67}$$

$$\overline{U}_2 = \frac{1}{\sqrt{4}} \begin{bmatrix} 1 \\ -1 \\ 1 \\ -1 \end{bmatrix} \tag{3.68}$$

$$\overline{U}_3 = \frac{1}{\sqrt{4}} \begin{bmatrix} 1 \\ 1 \\ -1 \\ -1 \end{bmatrix} \tag{3.69}$$

$$\overline{U}_4 = \frac{1}{\sqrt{4}} \begin{bmatrix} 1 \\ -1 \\ -1 \\ 1 \end{bmatrix} \tag{3.70}$$

Applying the eigenvectors one at a time to the sidewall hybrid in Figure 3.2*a* gives the eigennetworks in Figure 3.9. These eigennetworks coincide with electric and magnetic walls in all combinations along the symmetry planes. The scattering matrix of the junction is now obtained by diagonalizing it using Eq. 42 in Chapter 1.

$$S_{11} = \frac{s_1 + s_2 + s_3 + s_4}{4} \tag{3.71}$$

$$S_{12} = \frac{s_1 - s_2 + s_3 - s_4}{4} \tag{3.72}$$

$$S_{13} = \frac{s_1 + s_2 - s_3 - s_4}{4} \tag{3.73}$$

$$S_{14} = \frac{s_1 - s_2 - s_3 + s_4}{4} \tag{3.74}$$

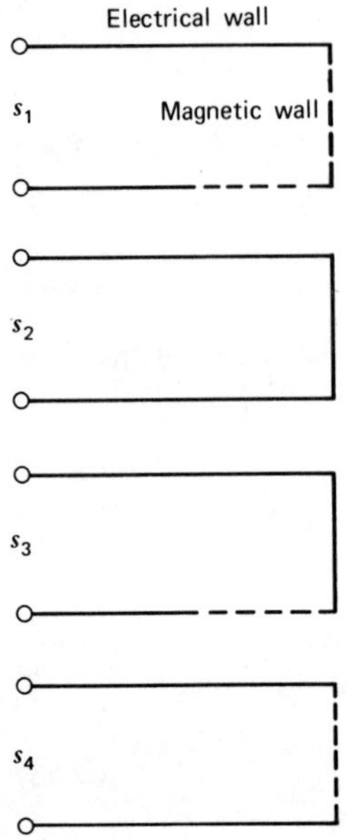

Figure 3.9. Eigennetworks of sidewall hybrid.

Inspection of the eigennetworks in Figure 3.19 or using the eigenvalues in Eqs. 3.63 through 3.66 immediately gives

$$S_{11}=0 \tag{3.75}$$

$$S_{12}=0 \tag{3.76}$$

$$S_{13}=\frac{s_1+s_2}{2}=\frac{1}{\sqrt{2}} \tag{3.77}$$

$$S_{14}=\frac{s_1-s_2}{2}=j\frac{1}{\sqrt{2}} \tag{3.78}$$

in agreement with Eqs. 3.16 and 3.39 through 3.42.

The correspondence between the odd and even modes and eigenvalue descriptions of the symmetrical directional coupler is completed by observing that the eigennetworks in Figures 3.9*b* and 3.9*d* support TE_{10} mode propagation, whereas those in Figures 3.9*a* and 3.9*c* support TE_{20} mode propagation.

3.6 *THE MULTIBRANCH DIRECTIONAL COUPLER*

The bandwidth of directional couplers may be enlarged in a number of different ways by using periodic coupling between the transmission lines.

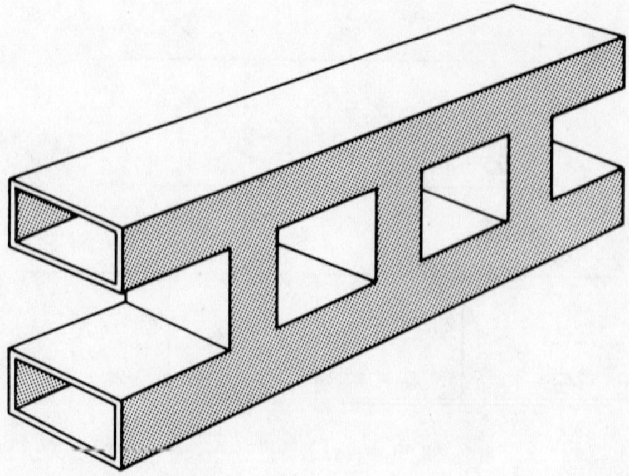

Figure 3.10. *Schematic diagram of waveguide multibranch coupler.*

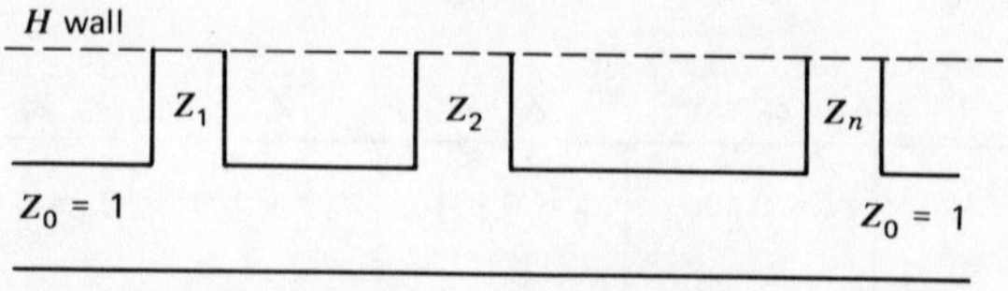

Figure 3.11a. *Even mode circuit for multibranch coupler.*

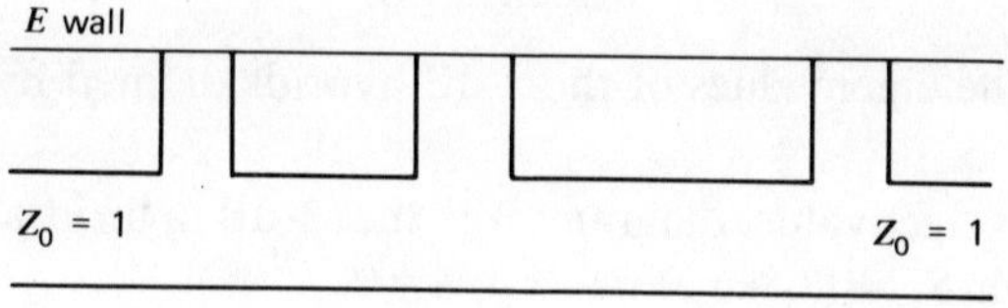

Figure 3.11b. *Odd mode circuit for multibranch coupler.*

One mechanical waveguide arrangement is illustrated in Figure 3.10. Odd and even mode circuits suitable for analysis of this structure are depicted in Figures 3.11*a* and 3.11*b*. Since these two circuits are ladder ones of the type encountered in the theory of filters, their branch immittances are closely related to that of filter structures. Multibranch coupling configurations based on a binomial coupling distribution are given in Figure 3.12.

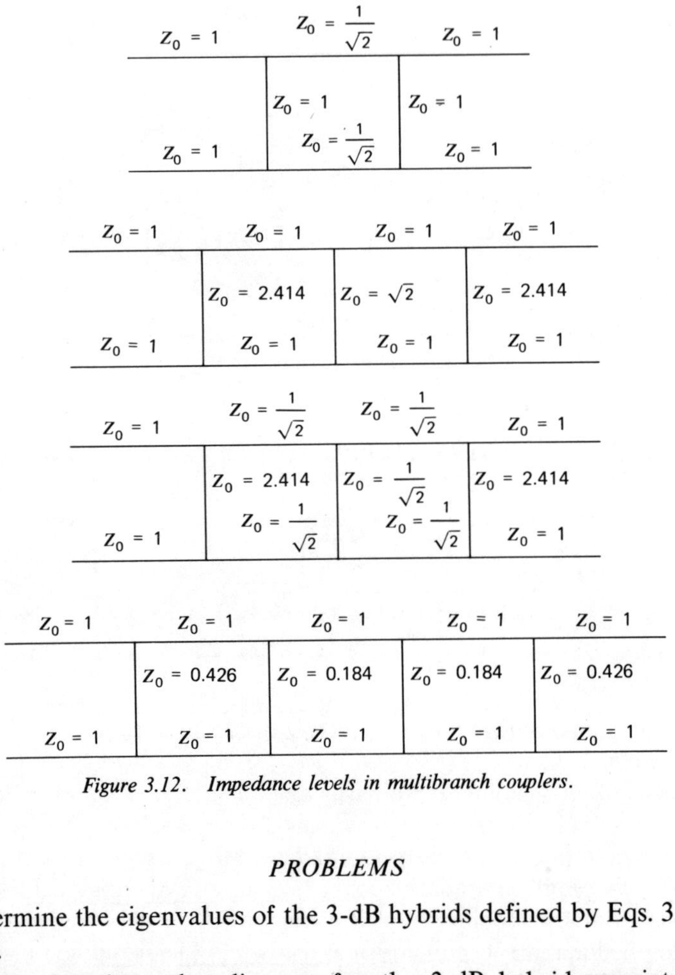

Figure 3.12. Impedance levels in multibranch couplers.

PROBLEMS

1. Determine the eigenvalues of the 3-dB hybrids defined by Eqs. 3.17 and 3.20.
2. Obtain the eigenvalue diagram for the 3-dB hybrid associated with the condition $S_{11}=(s_1+s_2+s_3+s_4)/4=0$.
3. Show that the scattering matrix for the magic-tee hybrid in Eq. 3.20 satisfies the unitary condition.
4. Obtain the outgoing waves for a 3-dB hybrid terminated at ports 3 and 4 in ganged short circuits with a single input at port 1.
5. Demonstrate that quadrature inputs at ports 3 and 4 of the magic tee produces similar outputs at ports 1 and 2.
6. Obtain the coupling length of the top wall 3-dB hybrid at 9 GHz for $a=22.8$ mm, $b=10.18$ mm, assuming that the odd and even mode propagation constants are TEM and TE_{10}.

7. Calculate the length of a 3-dB sidewall directional coupler at 9 GHz using waveguides with internal dimensions $a=22.8$ mm and $b=10.18$ mm.
8. By adding the scattered backward and forward waves of a two-hole coupler, show that the backscatter is minimized and the forwardscatter maximized when they are spaced a quarter wave apart.
9. A directional coupler with a directivity of 45 dB is used to measure a network with a return loss of 20 dB. Obtain the ripple on the return loss due to the finite directivity.

FURTHER READING

1. J. Paterson, A double slot hybrid junction, *L'Onde Electrique*, Special Supplement, Proceedings of the International Congress on Ultra High Frequency Circuits and Antennas, October 21-26, 1957.
2. R. Levy, Directional couplers, *Advance in Microwaves*, Vol. 1, L. Young, Ed., Academic Press, New York, 1966.
3. H. J. Riblet, *Proc. IRE*, **40** (1952), p. 180.

CHAPTER FOUR

Impedance and Mode Transducers

Microwave engineering involves many different types of transmission lines each characterized by its own mode of propagation and impedance level. A typical system usually involves more than one kind of line so that much of microwave engineering is concerned with the development of transitions between these different modes and impedance levels. Figure 4.1 illustrates just a few types of transmission lines with their dominant electric field patterns. This chapter briefly describes the theory of single and multistep impedance transformers. It also includes descriptions of mode transducers, rotary joints, and matched terminations. A feature of many of these components is their reliance on quarter wave long structures.

4.1 *QUARTER-WAVE IMPEDANCE TRANSFORMER*

One important class of matching network between transmission lines having different impedance levels is the quarter wave impedance transformer. The principle of this transformer is readily obtained by starting with the scattering matrix of a 2-port network terminated in a reflection coefficient S_L.

$$S'_{11} = S_{11} + \frac{S_{21}^2 S_L}{1 - S_{11} S_L} \tag{4.1}$$

This equation is obtained by setting a_2 equal to $S_L b_2$ in forming b_1/a_1, where S_L is given by

$$S_L = \frac{Z_L - Z_0}{Z_L + Z_0} \tag{4.2}$$

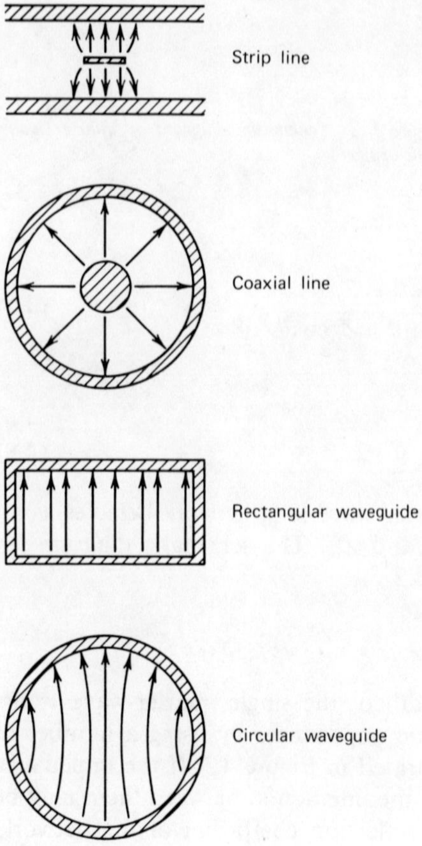

Figure 4.1. Typical transmission line configurations.

For a uniform lossless transmission line

$$S_{11}=0 \tag{4.3}$$

$$S_{21}=\exp(-j\theta) \tag{4.4}$$

Thus at the input terminals of the 2-port network S'_{11} becomes

$$S'_{11}=S_L\exp(-j2\theta) \tag{4.5}$$

The input admittance of the network is now given by

$$\frac{Z_{\text{in}}}{Z_0}=\frac{1+S'_{11}}{1-S'_{11}} \tag{4.6}$$

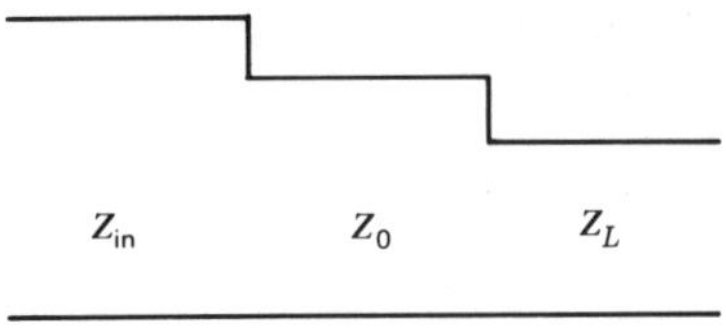

Figure 4.2. Schematic diagram of quarter wave transformer.

The result is

$$\frac{Z_{in}}{Z_0} = \frac{Z_L \cos\theta + jZ_0 \sin\theta}{jZ_L \sin\theta + Z_0 \cos\theta} \tag{4.7}$$

Putting $\theta = \pi/2$ gives

$$Z_{in} Z_L = Z_0^2 \tag{4.8}$$

This result indicates that such a quarter wave long network behaves as an impedance transformer between Z_L and Z_{in}. The schematic diagram for this arrangement is shown in Figure 4.2.

4.2 *STEPPED-IMPEDANCE TRANSFORMERS*

It is possible to extend the bandwidth of the single quarter-wave transformer between unequal characteristic impedances by using a number of them in cascade in the manner illustrated in Figure 4.3. If the impedance steps between the sections are small, the interaction between them may be neglected in calculating the overall reflection coefficient of the network. Using the nomenclature indicated in Figure 4.3, the overall reflection coefficient of the transformer may approximately be written as

$$S = S_1 + S_2 \exp(-j2\theta) + S_3 \exp(-j4\theta) + \ldots + S_n \exp[-j2(n-1)\theta] \tag{4.9}$$

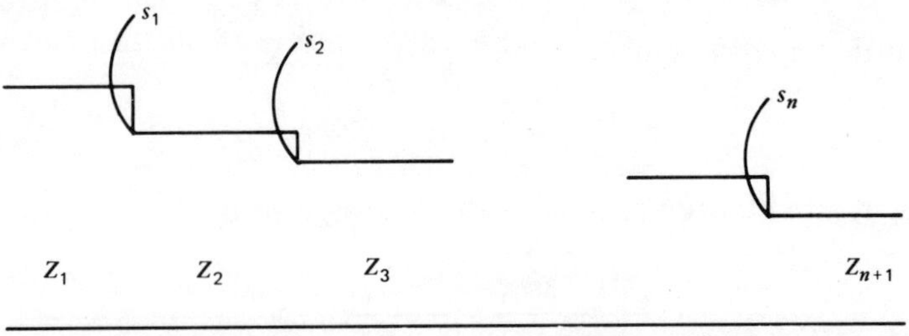

Figure 4.3. Schematic diagram of multistep transformer.

where the reflection coefficient of the ith step has the standard form given by

$$S_i = \frac{Z_{i+1} - Z_i}{Z_{i+1} + Z_i} \tag{4.10}$$

In what follows it is convenient to refer the overall reflection coefficient to the center of the transformer instead of at the input terminals

$$\begin{aligned} S_{11} &= S \exp[j(n-1)\theta] \\ &= S_1 \exp[j(n-1)\theta] + S_2 \exp[j(n-3)\theta] + \ldots + S_n \exp[-j(n-1)\theta] \end{aligned} \tag{4.11}$$

If the impedance steps are made symmetrical about the center of the transformer $S_1 = S_n$, $S_2 = S_{n-1}$, and so on, and the above equation becomes

$$S_{11} = S_1 \cos(n-1)\theta + S_2 \cos(n-3)\theta + S_3 \cos(n-5)\theta + \ldots + S_{n/2} \cos\theta \quad \text{for } n \text{ even} \tag{4.12}$$

$$S_{11} = S_1 \cos(n-1)\theta + S_2 \cos(n-3)\theta + \ldots + S_{(n-1)/2} \cos 2\theta + S_{(n+1)/2} \quad \text{for } n \text{ odd} \tag{4.13}$$

The reflection coefficient of the transformer may now be optimized by choosing appropriate values for S_i. The two classic arrangements are obtained by having the reflection coefficients coincide with either a binomial or Chebyshev polynomial.

Choosing the Chebyshev polynomial as an example and formulating S_{11} in terms of S_{21} gives

$$|S_{11}|^2 = 1 - |S_{21}|^2 \tag{4.14}$$

where

$$|S_{21}|^2 = \frac{1}{1 + \epsilon^2 T_{n-1}^2(x)} \tag{4.15}$$

Thus

$$S_{11} \simeq \epsilon T_{n-1}(x) \tag{4.16}$$

where ϵ is the ripple level defined in Figure 4.4 and $T_n(x)$ is the Chebyshev

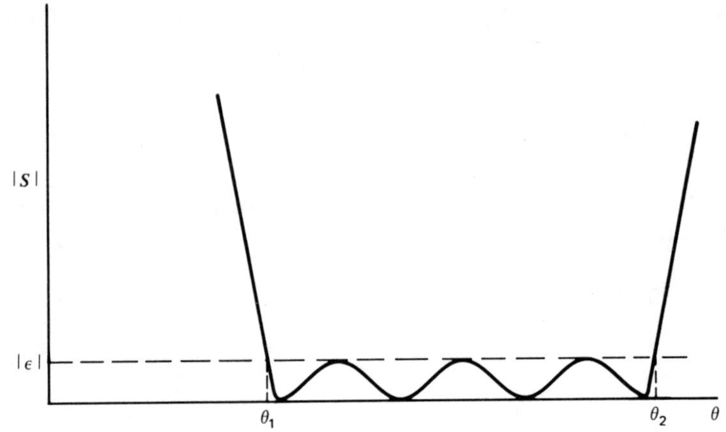

Figure 4.4. Chebyshev response for n = 5 *network.*

function defined by

$$T_n(x) = \cos(n\cos^{-1}x) \qquad |x| \leqq 1 \tag{4.17}$$

$$T_n(x) = \cosh(n\cosh^{-1}x) \qquad |x| \geqq 1 \tag{4.18}$$

The first few terms in this series are

$$T_0(x) = 1 \tag{4.19}$$

$$T_1(x) = x \tag{4.20}$$

$$T_2(x) = 2x^2 - 1 \tag{4.21}$$

$$T_3(x) = 4x^3 - 3x \tag{4.22}$$

and

$$T_{n+1}(x) = 2xT_n(x) - T_{n-1}(x) \tag{4.23}$$

where

$$x = \frac{\cos\theta}{\cos\theta_1} \tag{4.24}$$

The frequency variables θ and θ_1 are defined as

$$\theta = \frac{\pi}{2}(1+\delta) \tag{4.25}$$

$$\theta_1 = \frac{\pi}{2}(1+\delta_1) \tag{4.26}$$

where

$$\delta = \frac{\omega - \omega_0}{\omega_0} \tag{4.27}$$

$$\delta_1 = \frac{\omega_1 - \omega_0}{\omega_0} \tag{4.28}$$

ω_0 is the center frequency of the transformer, $\omega_{1,2}$ the band edges, and ω is the normal frequency variable.

The design now proceeds by writing Eqs. 4.12 and 4.13 as polynomials in x and equating like coefficients in Eq. 4.16. This is done by first replacing $\cos\theta$ by $\cos\theta_1$ in Eqs. 4.12 and 4.13 and then introducing the following trigonometric identities

$$\cos\theta = x\cos\theta_1 \tag{4.29}$$

$$\cos 2\theta = 2x^2\cos^2\theta_1 - 1 \tag{4.30}$$

$$\cos 3\theta = 4x^3\cos^3\theta_1 - 3x\cos\theta_1 \tag{4.31}$$

and so on.

Equation 4.16 gives a relation between θ_1, n, ϵ, and the reflection coefficients of the steps. However, an additional relation must be constructed to ensure that the two immittances are correctly joined. This relation is obtained by taking a linear combination of the S's with $\theta = 0$.

$$\sum_{i=1}^{n+1} S_i = \tfrac{1}{2}\ln\left(\frac{Z_{n+1}}{Z_1}\right) \tag{4.32}$$

In obtaining this relation the following approximation is used to represent S_i in Eq. 4.32.

$$S_i = \frac{Z_{i+1} - Z_i}{Z_{i+1} + Z_i} = \tfrac{1}{2}\ln\left(\frac{Z_{i+1}}{Z_i}\right) = \tfrac{1}{2}(\ln Z_{i+1} - \ln Z_i) \tag{4.33}$$

Taking $n = 3$ as an example, Eqs. 4.13 and 4.16 gives

$$2S_1(2x^2\cos^2\theta_1 - 1) + S_2 = \epsilon(2x^2 - 1) \tag{4.34}$$

Thus

$$S_1 = \frac{\epsilon}{2\cos^2\theta_1} \tag{4.35}$$

$$S_2 = \epsilon\left(\frac{1}{\cos^2\theta_1} - 1\right) \tag{4.36}$$

Satisfying Eq. 4.32 gives

$$(2S_1+S_2)=\tfrac{1}{2}\ln\left(\frac{Z_4}{Z_1}\right) \tag{4.37}$$

The ripple level is therefore given in terms of the bandwidth parameter θ_1 and Z_4/Z_1 by

$$\epsilon\left(\frac{2}{\cos\theta_1^2}-1\right)=\tfrac{1}{2}\ln\left(\frac{Z_4}{Z_1}\right) \tag{4.38}$$

If this value of ϵ is inadequate, the procedure is repeated with $n=4$.

The impedances of the steps are finally given in terms of S_1 and S_2 by

$$S_1=S_3=\tfrac{1}{2}\ln\left(\frac{Z_2}{Z_1}\right)=\tfrac{1}{2}\ln\left(\frac{Z_4}{Z_3}\right) \tag{4.39}$$

$$S_2=\tfrac{1}{2}\ln\left(\frac{Z_3}{Z_2}\right) \tag{4.40}$$

As an example consider the design of a two-section ($n=3$) transformer between transmission lines of impedance 16.66 Ω (Z_1) and 50 Ω (Z_4) over a bandwidth of 60%.

For this bandwidth specification θ_1 is given from Eq. 4.26 as

$$\theta_1=63\text{ deg}$$

Using this value of θ_1 in Eq. 4.38 gives the ripple level as

$$\epsilon=0.063$$

Once θ_1 and ϵ are stated the three impedance steps are given by Eqs. 4.35 and 4.36 by

$$S_1=S_3=0.153$$

$$S_2=0.243$$

Finally the impedance levels of the two transformer sections are given from Eq. 4.39 by

$$Z_2=Z_1\text{ anti ln }2S_1=22.62\ \Omega$$

$$Z_3=\frac{Z_4}{\text{anti ln }2S_1}=36.82\ \Omega$$

It is observed that the ratio of Z_3/Z_2 is consistent with Eq. 4.40.

4.3 *MATCHED TERMINATIONS*

It is often necessary in microwave engineering to terminate a transmission line in its characteristic impedance. One coaxial version consists of a dissipative resistive film accross the transmission line with a shortcircuit piston behind it in the manner indicated in Figure 4.5. The operation of this load is readily understood by forming its reflection coefficient.

$$S=\frac{Y_0-Y_{\text{in}}}{Y_0+Y_{\text{in}}} \tag{4.41}$$

where

$$Y_{\text{in}}=G+jY_0\cot\theta \tag{4.42}$$

Setting the reflection coefficient to zero to obtain a matched termination gives

$$G=Y_0 \tag{4.43}$$

$$Y_0\cot\theta=0 \tag{4.44}$$

This gives $\theta=\pi/2$, which coincides with a quarter-wave-long section between the planes of the shortcircuit and resistive film.

Figure 4.6 depicts one waveguide-matched load consisting of a tapered section of dielectric material loaded with a lossy polyiron material that is

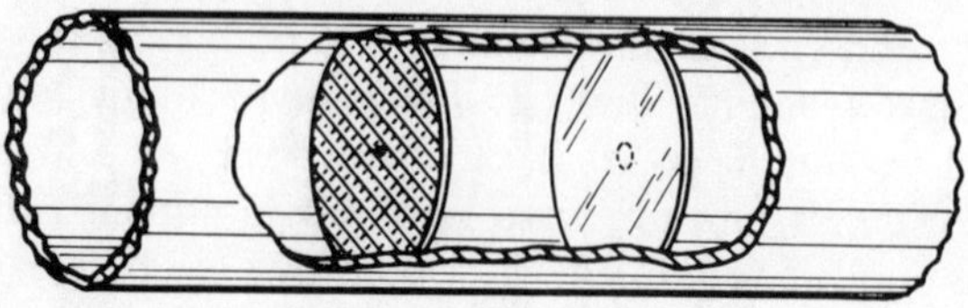

Figure 4.5. Construction of coaxial termination using resistive film.

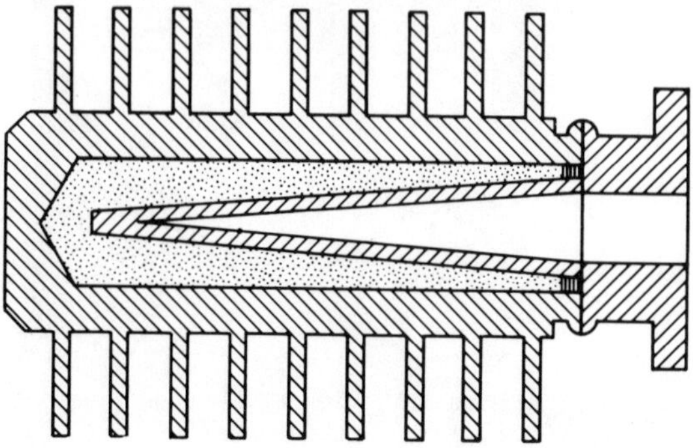

Figure 4.6. Construction of waveguide termination.

gradually introduced into the waveguide. If the taper is sufficiently smooth it will attenuate the microwave signal without any reflections.

4.4 *ROTARY JOINT*

Microwave rotary joints are employed to link the stationary part with the rotating part of the radar. It relies for its operation on a section of transmission line supporting a circularly symmetrical mode of propagation with the conductors suitably cut to permit relative movement. The most common modes employed in the construction of these devices are the coaxial TEM mode and the circular waveguide TM_{01} one in Figure 4.7. The breaks in the transmission lines are electrically closed with noncontacting chokes or sliding contacts.

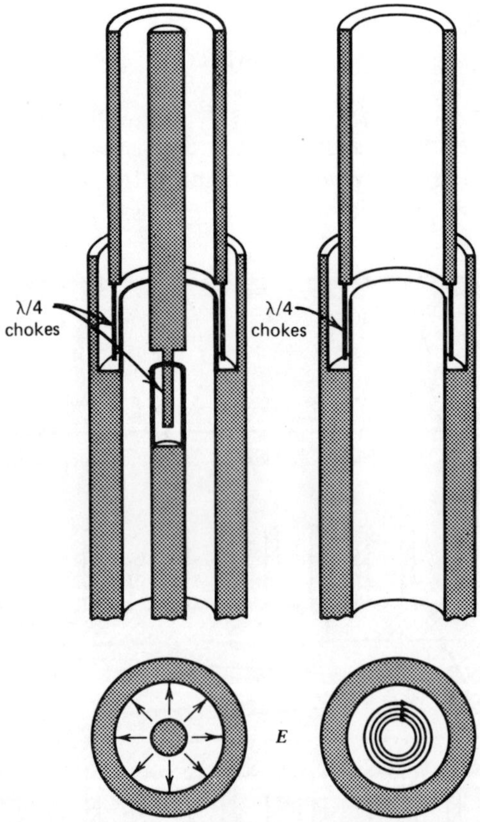

Figure 4.7. TEM and TM_{01} circular symmetric modes.

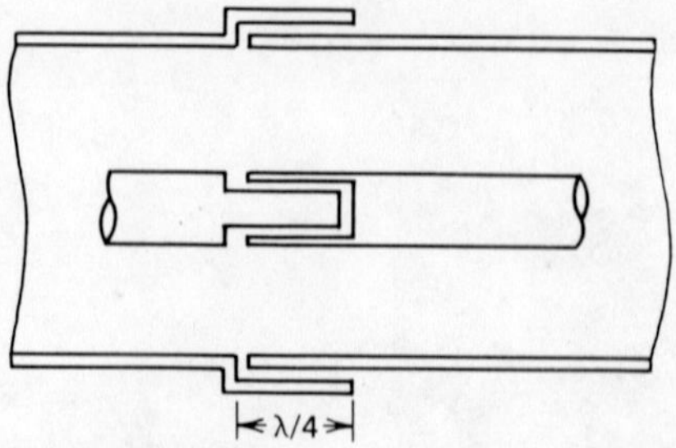

Figure 4.8a. Single quarter-wave choke construction.

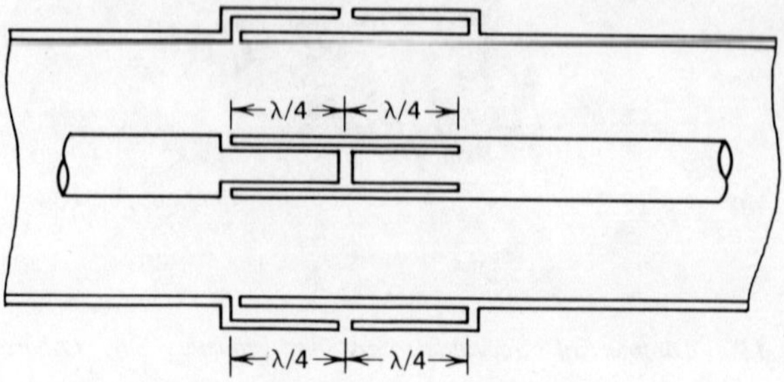

Figure 4.8b. Double quarter-wave choke construction.

In Figure 4.8*a* choke sections are placed in series with the gaps between the fixed and rotating parts of the joint so that a low impedance is located at the input terminals of the two gaps. Forming the input impedance of the choke section by letting $Z_L = \infty$ in Eq. 4.7 yields

$$Z_{\text{in}} = jZ_0 \tan\theta \tag{4.45}$$

This gives

$$Z_{\text{in}} = 0 \qquad \text{for}\ \theta = \pi/2 \tag{4.46}$$

which corresponds to a $\lambda/4$ transmission line.

Since such a choke would normally be frequency sensitive, the 2-section choke geometry illustrated in Figure 4.8*b* is often used instead. This $\lambda/2$ transmission line transforms the shortcircuit termination to the input terminals of the gaps. As the impedance is an open circuit midway along the choke, the transmission lines may be cut at these points to allow

Figure 4.9. Commercial waveguide rotary joint (courtesy Sage Laboratories).

rotation of the two halves without perturbing propagation along the main transmission line. Figure 4.9 shows one commercial rotary joint.

4.5 *MODE TRANSDUCERS*

Since the modes of transmission in microwave engineering are not unique, it is often necessary to construct transitions between them. Figure 4.1 illustrates one line, while Figure 4.10 depicts some higher order modes in circular waveguide. Transitions between waveguides operating in their dominant modes usually consist either of linear tapers or quarter wave sections of intermediate cross section to those being joined. Coupling to higher order modes involves the use of a number of probes or loops suitably spaced around the transmission line with the symmetry of the higher order mode.

Figure 4.11 indicates one important practical transition between a rectangular waveguide and a coaxial transmission line. The shortcircuit piston reduces an otherwise 3-port junction to a 2-port device. It is observed that both the length of the coaxial probe and the shortcircuit section are approximately a quarter wave long.

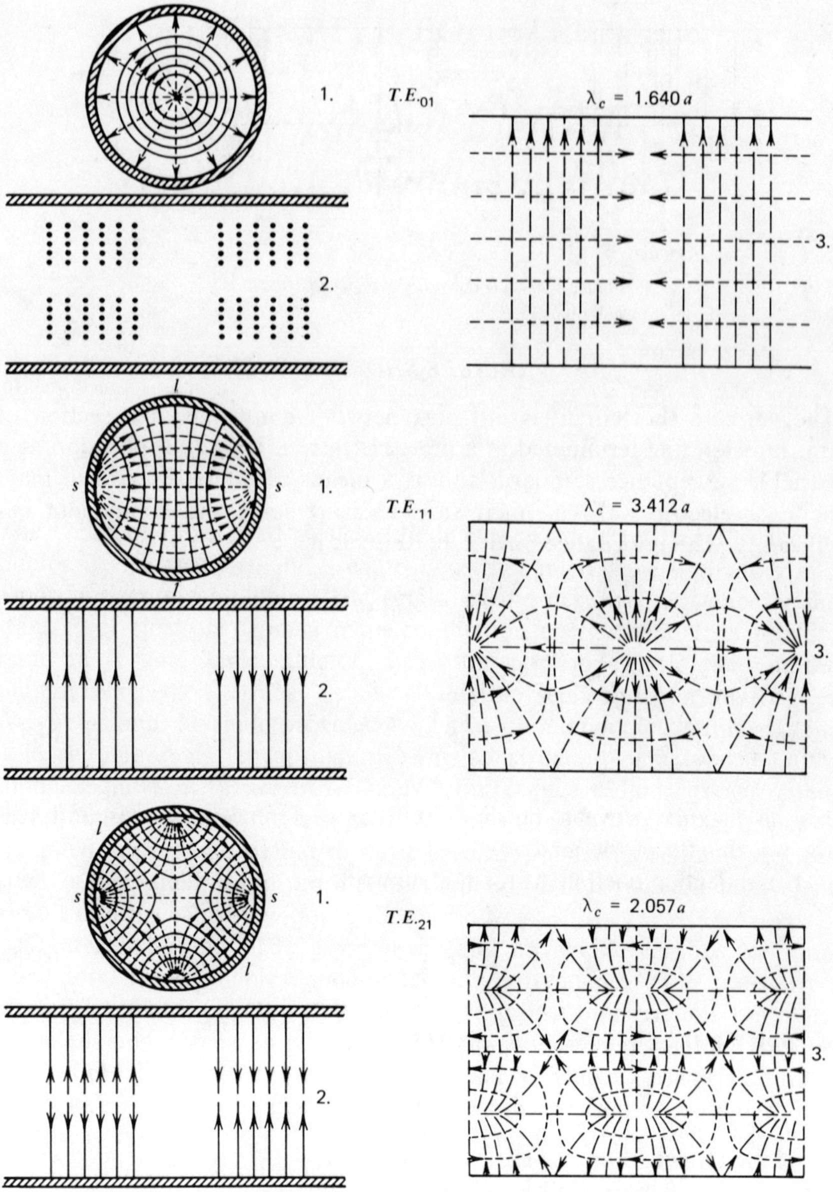

Field distribution for H—modes in circular waveguide.

1. Cross-sectional view
2. Longitudinal view through plane l–l
3. Surface view from s–s

Figure 4.10. Higher order modes in circular waveguide.

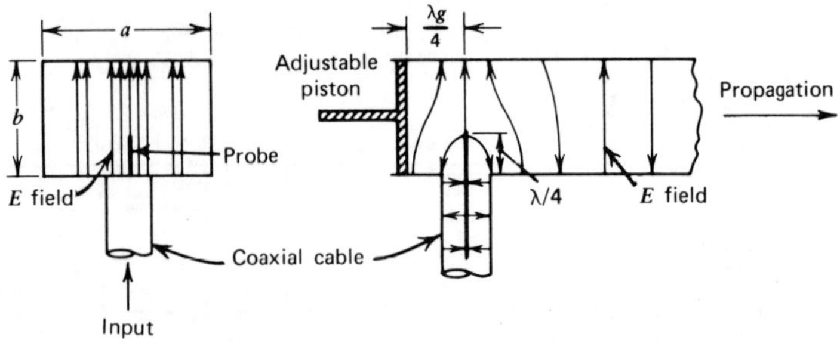

Figure 4.11. Waveguide to coaxial transition.

4.6 *VARIABLE SHORTCIRCUIT*

The variable shortcircuit is a 1-port network consisting of a section of transmission line terminated in a movable piston. It finds application as a variable susceptance standard, and as a means of establishing ideal magnetic or electric walls in microwave measurements and component engineering. The impedance of the shortcircuit is

$$Z_{\text{in}} = jZ_0 \tan\theta \tag{4.47}$$

where

$$\theta = \frac{2\pi l}{\lambda_g} \tag{4.48}$$

l is the distance between the shortcircuit and reference terminals, and λ_g is the wavelength of the medium.

The reflection coefficient for this network is

$$S_{11} = \frac{Z_{\text{in}} - Z_0}{Z_{\text{in}} + Z_0} \tag{4.49}$$

Combining the above equations gives

$$S_{11} = -\exp(-j2\theta) \tag{4.50}$$

Thus the reflection coefficient exhibits a unity reflection coefficient with a variable phase angle. This is just the form adopted in Chapter 5.

To ensure that the reflection coefficient is nearly unity, the movable shortcircuit is usually fitted with spring fingers, or with a choke arrangement similar to that used in the design of the rotary joint. This is shown in Figure 4.12.

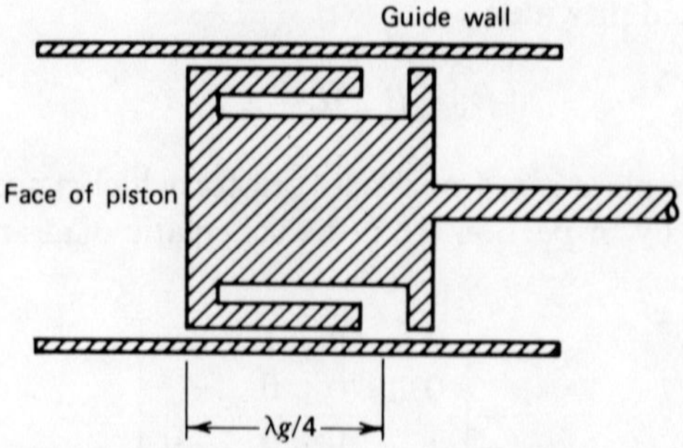

Figure 4.12. Variable short circuit piston in waveguide.

4.7 QUARTER WAVE PLATE

The notion of horizontal and vertical linear polarization is well established. However, such polarizations may be regarded as limiting cases of elliptical polarization. Clockwise and anticlockwise polarizations that may also be defined are the other limiting forms of elliptical polarization. Many systems and microwave components in fact rely on the use of such circular polarization. To convert a linearly polarized wave into a circularly polarized one it is necessary to decompose the incoming wave into two orthogonal components with equal amplitudes with one of them phase shifted through $\pm\pi/4$ rad with respect to the other.

One common type of circular polarizer or quarter-wave plate consists of a sheet of dielectric material at $\pm\pi/4$ rad to the x and y axis of a circular waveguide; in the manner indicated in Figure 4.13. If a vertically polarized wave is introduced into such a waveguide, it will have equal components perpendicular and parallel to the dielectric sheet that propagate with different velocities $\beta\|$ and $\beta\perp$ along the waveguide. The component perpendicular to the dielectric sheet remains essentially unperturbed, whereas the component parallel to it is increased. The output wave will be

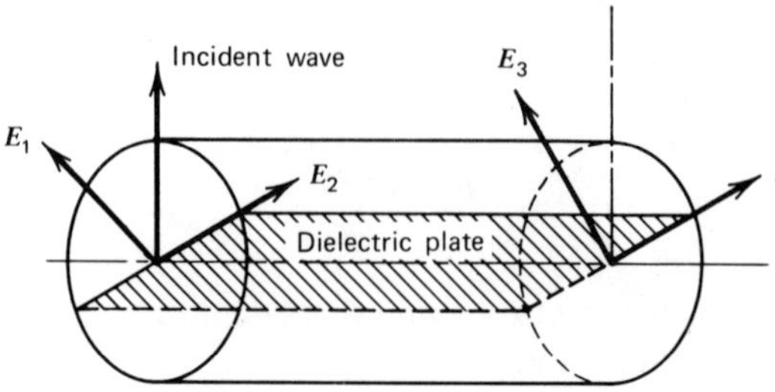

Figure 4.13. Quarter-wave plate.

circularly polarized provided

$$(\beta\| - \beta\perp)L = \pm\pi/4 \tag{4.51}$$

The quarter-wave plate is a 4-port network for which a scattering matrix is readily obtained by inspection from the schematic diagram in Figure 4.13. The result is

$$\bar{S} = \begin{bmatrix} 0 & 0 & 1 & 0 \\ 0 & 0 & 0 & -j \\ 1 & 0 & 0 & 0 \\ 0 & -j & 0 & 0 \end{bmatrix} \tag{4.52}$$

In obtaining this matrix it has been assumed that the network is matched so that the main diagonal is zero, that it is reciprocal so that the matrix is symmetrical about the main diagonal, that all ports are decoupled except ports 1 and 3 and 2 and 4, and that there is a $\pi/4$ differential phase shift between ports 3 and 4.

If a linearly polarized wave is now incident on the device with its polarization at 45° to ports 1 and 2 the outgoing waves are obtained by forming the following input/output relation:

$$\begin{bmatrix} b_1 \\ b_2 \\ b_3 \\ b_4 \end{bmatrix} = \begin{bmatrix} 0 & 0 & 1 & 0 \\ 0 & 0 & 0 & -j \\ 1 & 0 & 0 & 0 \\ 0 & -j & 0 & 0 \end{bmatrix} \begin{bmatrix} \frac{1}{\sqrt{2}} \\ \frac{1}{\sqrt{2}} \\ 0 \\ 0 \end{bmatrix} \tag{4.53 }$$

The result is

$$b_1 = 0 \tag{4.54}$$

$$b_2 = 0 \tag{4.55}$$

$$b_3 = \frac{1}{\sqrt{2}} \tag{4.56}$$

$$b_4 = \frac{-j}{\sqrt{2}} \tag{4.57}$$

Thus the outgoing wave at ports 3 and 4 is circularly polarized, in that it is in space and time quadrature.

If the dielectric sheet is rotated by 90 deg. with respect to its position in Figure 4.13, S_{13} and S_{14} are interchanged and the input/output relation of the network becomes

$$\begin{bmatrix} b_1 \\ \\ b_2 \\ \\ b_3 \\ b_4 \end{bmatrix} = \begin{bmatrix} 0 & 0 & -j & 0 \\ \\ 0 & 0 & 0 & 1 \\ \\ -j & 0 & 0 & 0 \\ 0 & 1 & 0 & 0 \end{bmatrix} \begin{bmatrix} \frac{1}{\sqrt{2}} \\ \frac{1}{\sqrt{2}} \\ 0 \\ 0 \end{bmatrix} \tag{4.58}$$

The result is

$$b_1 = 0 \tag{4.59}$$

$$b_2 = 0 \tag{4.60}$$

$$b_3 = \frac{-j}{\sqrt{2}} \tag{4.61}$$

$$b_4 = \frac{1}{\sqrt{2}} \tag{4.62}$$

Thus the output wave at ports 3 and 4 is now circularly polarized in the opposite direction. If two similar quarter-wave plates are cascaded, a vertically polarized wave incident on the first plate is converted to an horizontal one at the output plate. However, if one plate of each type are connected in tandem, the polarization at the output plate is the same as that at the input.

PROBLEMS

1. Show that the matrix given by Eq. 4.52 satisfies the unitary condition.
2. Show that a circularly polarized wave is converted into a linearly polarized one in passing through a quarter-wave plate.
3. Obtain the input impedance Z_{in} of a two-section impedance transformer terminated in a load Z_L.
4. Show that for an $n=2$ binomial transformer $Z_1 = Z_L^{1/4} Z_0^{3/4}$, $Z_2 = Z_L^{3/4} Z_0^{1/4}$.
5. Design a two-section transformer to match a transmission line of 50 Ω characteristic impedance to one of 5 Ω over a 10% band.
6. Verify that Eqs. (19) through (22) satisfy Eq. (23).

FURTHER READING

1. S.B. Cohn, Optimum design of stepped transmission line transformers, *IRE Trans. (on MTT*, April 1955, pp. 16–21.
2. J.L. Altman, *Microwave Circuits*, Van Nostrand, New York, 1964.

CHAPTER FIVE

Two-Port Phase and Attenuation Networks

The ideal form of phase circuit is a 2-port network having a transfer function of unit amplitude and mechanically or electronically variable phase. This definition of the phase shifter is expressed by the following scattering matrix

$$\bar{S}=\begin{bmatrix} 0 & e^{-j\theta} \\ e^{-j\theta} & 0 \end{bmatrix} \tag{5.1}$$

The ideal form of a variable attenuator is a 2-port network having zero phase shift and a mechanically or electronically variable attenuation. The scattering matrix for this network is

$$\bar{S}=\begin{bmatrix} 0 & e^{-\alpha} \\ e^{-\alpha} & 0 \end{bmatrix} \tag{5.2}$$

These two definitions assume that both networks are reciprocal and perfectly matched. However, as discussed elsewhere, nonreciprocal phase shifters and attenuators may be realized by having recourse to ferrite devices. In what follows three different configurations of each function will be discussed. The circuits described are uncalibrated waveguide devices suitable for phase and amplitude trimming, high-power circuits employing hybrid networks, and phase and attenuation standards using resistive and dielectric rotors in circular waveguides.

5.1 *UNCALIBRATED WAVEGUIDE PHASE-SHIFTERS AND ATTENUATORS*

Simple mechanical types of waveguide phase shifters and attenuators consist of either a movable dielectric slab or resistive sheet across a waveguide. Figures 5.1 and 5.2 illustrates usual mechanical arrangements. The dielectric slab produces a phase delay that is dependent on its position across the waveguide. The attenuation of the resistive card is also a function of its location in the waveguide.

The two structures can be analyzed, if necessary, by establishing the transverse resonance condition across each waveguide configuration. However, it is seldom necessary to calibrate these components. The analysis of these devices proceeds by forming the ABCD matrix for each of the three regions, multiplying them out to form the overall matrix, and applying the boundary conditions at the sides of the waveguide. This gives a transcendental equation from which either θ or α can be obtained for use in Eqs. 5.1 and 5.2.

The overall ABCD is defined by

$$\begin{bmatrix} A & jB \\ jC & D \end{bmatrix} = \begin{bmatrix} A_1 & jB_1 \\ jC_1 & D_1 \end{bmatrix} \times \begin{bmatrix} A_2 & jB_2 \\ jC_2 & D_2 \end{bmatrix} \times \begin{bmatrix} A_3 & jB_3 \\ jC_3 & D_3 \end{bmatrix} \quad (5.3)$$

where ABCD relates the fields at the two waveguide walls

$$\begin{bmatrix} E_{y1} \\ H_{x1} \end{bmatrix} = \begin{bmatrix} A & jB \\ jC & D \end{bmatrix} \begin{bmatrix} E_{y3} \\ H_{x3} \end{bmatrix} \quad (5.4)$$

For a reciprocal waveguide

$$A = D \quad (5.5)$$

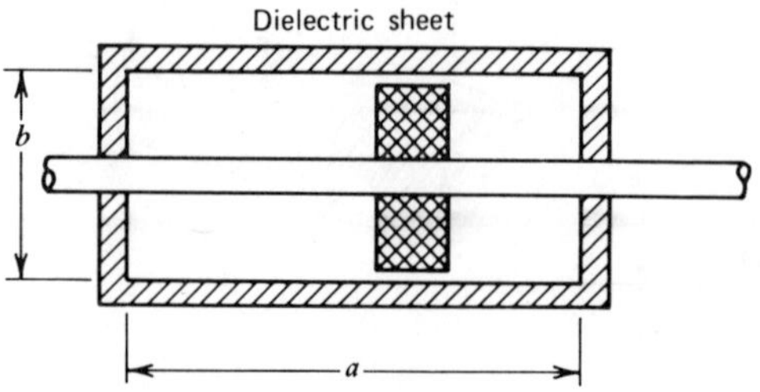

Figure 5.1. Section view of waveguide dielectric phase shifter.

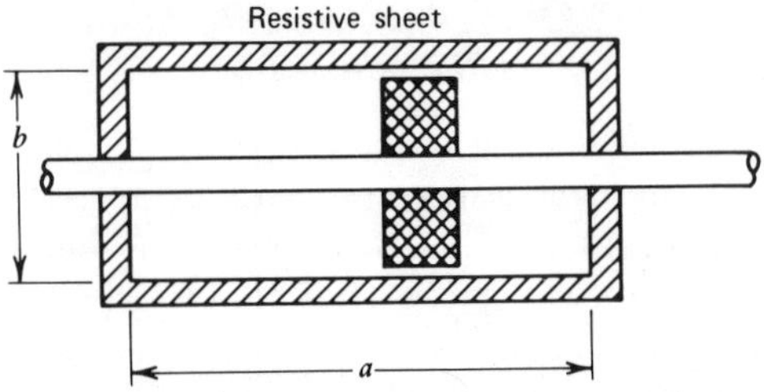

Figure 5.2. Section view of waveguide resistive card attenuator.

The boundary conditions at the sides of the waveguide are

$$E_{y1} = E_{y3} = 0 \tag{5.6}$$

Thus, the transcendental equation for θ or α is

$$B = 0 \tag{5.7}$$

Although the derivation of the ABCD matrices is straightforward it is outside the scope of this text.

5.2 *VARIABLE ROTOR ATTENUATOR*

The ideal rotary vane attenuator to be described now is a variable attenuation standard having a $\cos^2\theta$ attenuation law. It consists of three sections of waveguide in tandem in the manner indicated in Figure 5.3. The two end sections are rectangular to circular waveguide transitions, whereas the center section is a circular waveguide free to rotate between

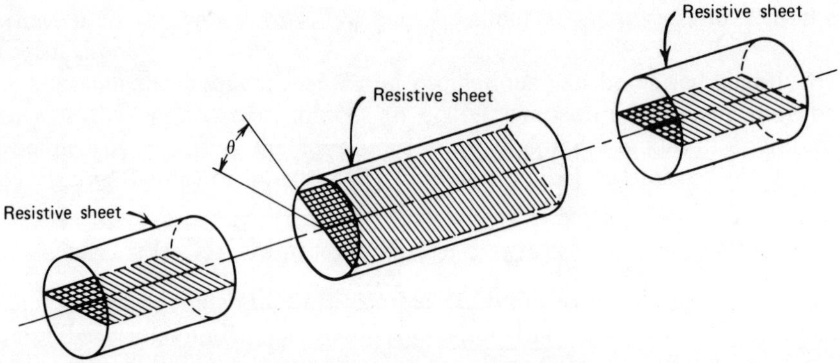

Figure 5.3. Rotary vane attenuator.

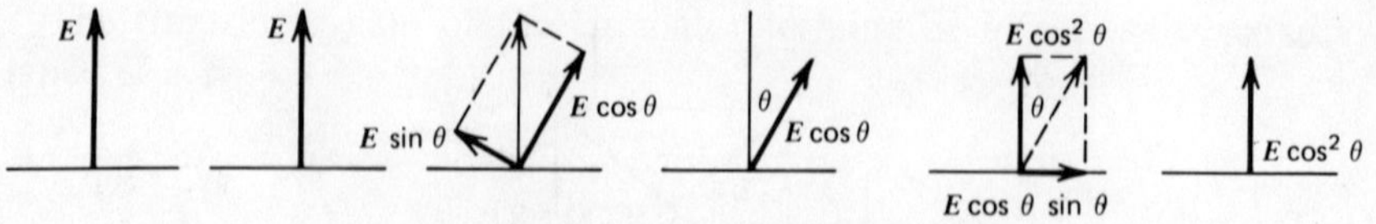

Figure 5.4. Rotary vane attenuator voltage vectors.

the stator sections. Across the center of each waveguide is a resistive card. The scattering matrix for this device is

$$\bar{S}=\begin{bmatrix} 0 & S_{21} \\ S_{21} & 0 \end{bmatrix} \tag{5.8}$$

where for an ideal device

$$S_{21}=\cos^2\theta \tag{5.9}$$

and θ is the angle between the resistive cards of the two stators and that of the rotor.

The form of S_{21} may be understood by considering an incoming wave perpendicular to the resistive card of the first stator section. If the resistive card of the rotor makes an angle θ with respect to those of the two stators, an incident wave on the rotor may be decomposed into components parallel and perpendicular to its resistive card given by $E_0\cos\theta$ and $E_0\sin\theta$. This is shown in Figure 5.4.

At the output of the rotor section, the wave parallel to the resistive card is completely attenuated by it, whereas that perpendicular to it is unattenuated. The latter component is now resolved into components perpendicular and parallel to the resistive card of the second stator. The parallel term is now absorbed by the last resistive card, whereas the perpendicular one $E_0\cos^2\theta$ becomes the outgoing wave, which is the desired result.

5.3 *HIGH-POWER ATTENUATOR*

The two attenuators described so far have been low-power ones in that they have both relied on dissipation in a resistive card suitably placed in a waveguide. One waveguide variable attenuator that is particularly attractive for high-power use is shown in Figure 5.5. It consists of a waveguide hybrid terminated in a variable shortcircuit in series with one of the two connecting lines of a cascade arrangement of two hybrids. In this circuit the high-power signal is dissipated in a high-power load, which is either air or water cooled.

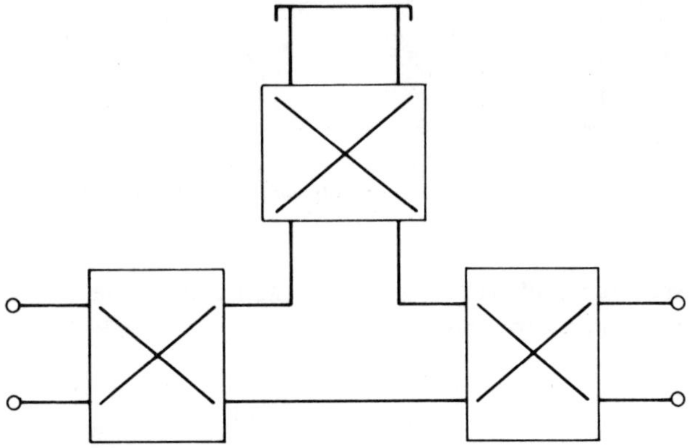

Figure 5.5. Schematic diagram of high-power attenuator using 3-dB hybrids.

The operation of this device may be understood by forming the input/output waves of each hybrid one at a time. The outgoing waves for the first hybrid for an input wave at port 1 are given in Chapter 3 by

$$b_1=0 \tag{5.10}$$

$$b_2=0 \tag{5.11}$$

$$b_3=\frac{1}{\sqrt{2}} \tag{5.12}$$

$$b_4=\frac{j}{\sqrt{2}} \tag{5.13}$$

b_3 is now phase shifted by the series hybrid through a phase angle θ_1, which is determined by the position of the variable shortcircuit, whereas b_4 is phase shifted through a fixed phase angle θ_2 determined by the series waveguide section connecting the two hybrids.

The incident waves at the input terminals of the output hybrid are therefore

$$a_1=\frac{1}{\sqrt{2}}e^{-j\theta_1} \tag{5.14}$$

$$a_2=\frac{j}{\sqrt{2}}e^{-j\theta_2} \tag{5.15}$$

The output waves of the output hybrid are now obtained by forming the input/output relation for that hybrid

$$\begin{bmatrix} b_1 \\ b_2 \\ b_3 \\ b_4 \end{bmatrix} = \frac{1}{\sqrt{2}} \begin{bmatrix} 0 & 0 & 1 & j \\ 0 & 0 & j & 1 \\ 1 & j & 0 & 0 \\ j & 1 & 0 & 0 \end{bmatrix} \begin{bmatrix} \frac{1}{\sqrt{2}} e^{-j\theta_1} \\ \frac{j}{\sqrt{2}} e^{-j\theta_2} \\ 0 \\ 0 \end{bmatrix} \tag{5.16}$$

Thus the outgoing waves of the last hybrid are

$$b_1 = 0 \tag{5.17}$$

$$b_2 = 0 \tag{5.18}$$

$$b_3 = \frac{e^{-j\theta_1} - e^{-j\theta_2}}{2} \tag{5.19}$$

$$b_4 = \frac{j(e^{-j\theta_1} + e^{-j\theta_2})}{2} \tag{5.20}$$

Setting $\theta_1 = \theta_2$ gives

$$b_3 = 0 \tag{5.21}$$

$$b_4 = j \tag{5.22}$$

If $\theta_1 = \theta_2 + \pi$ the result is

$$b_3 = 1 \tag{5.23}$$

$$b_4 = 0 \tag{5.24}$$

The arrangement in Figure 5.5 may therefore be used to divide an incident wave at port 1 between ports 3 and 4 by varying the position of the shortcircuit piston of the device. A physical realization of such a device is illustrated in Figure 5.6. In obtaining the above result only relative phase angles have been preserved.

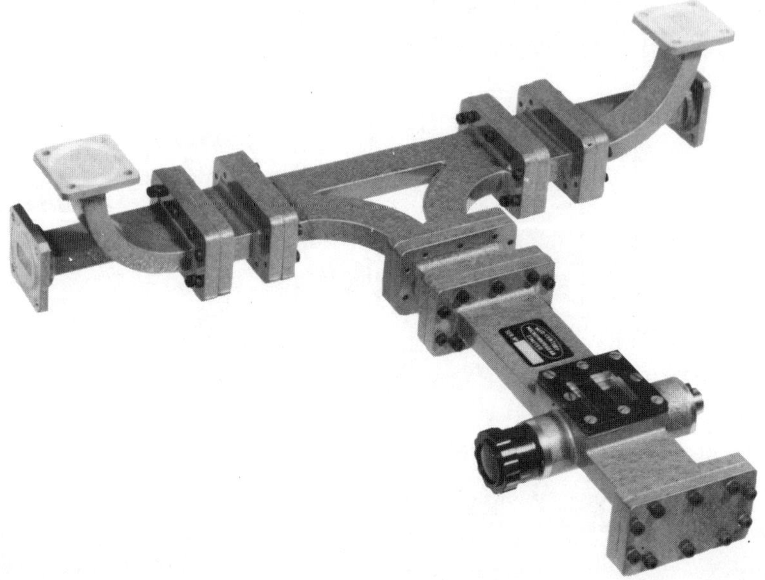

Figure 5.6. Commercial high power waveguide attenuator using 3-d B hybrids (courtesy Mid-Century Ltd).

5.4 *VARIABLE ROTOR PHASE SHIFTER*

A standard variable rotor phase shifter may be constructed by placing a half-wave plate between two similar quarter-wave ones in the manner depicted in Figure 5.7. The first quarter-wave plate converts a linearly polarized wave into a circularly polarized one that is phase shifted through an angle 2θ, whereas the second plate converts the latter wave back to linear polarization.

The operation of this device is readily understood by forming the input/output relation of each section one at a time through their respective scattering matrices. The half-wave plate resembles the quarter-wave one in construction but introduces a 180-deg phase delay instead of a 90-deg delay. Such a plate is shown in Figure 5.8. This plate will produce a continuous change in phase in a circularly polarized wave, if rotated about its axis, retarding or advancing it depending on the sense of rotation of the circular wave. It may therefore be combined with two quarter-wave plates to form a variable phase shifter. The scattering matrix of the half-wave

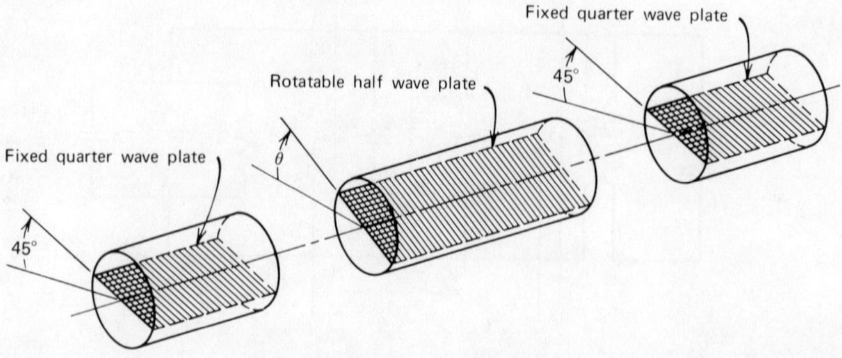

Figure 5.7. Rotary vane phase shifter.

plate may be shown to have the following form

$$\begin{bmatrix} 0 & 0 & -\cos 2\theta & -\sin 2\theta \\ 0 & 0 & -\sin 2\theta & \cos 2\theta \\ -\cos 2\theta & -\sin 2\theta & 0 & 0 \\ -\sin 2\theta & \cos 2\theta & 0 & 0 \end{bmatrix} \tag{5.25}$$

The behavior of the half-wave plate may be verified by taking the circularly polarized outgoing wave of the quarter-wave plate as its input and forming its input/output relation using the above scattering matrix.

$$\begin{bmatrix} b_1 \\ b_2 \\ b_3 \\ b_4 \end{bmatrix} = \begin{bmatrix} 0 & 0 & -\cos 2\theta & -\sin 2\theta \\ 0 & 0 & -\sin 2\theta & \cos 2\theta \\ -\cos 2\theta & -\sin 2\theta & 0 & 0 \\ -\sin 2\theta & \cos 2\theta & 0 & 0 \end{bmatrix} \begin{bmatrix} \frac{1}{\sqrt{2}} \\ \frac{-j}{\sqrt{2}} \\ 0 \\ 0 \end{bmatrix} \tag{5.26}$$

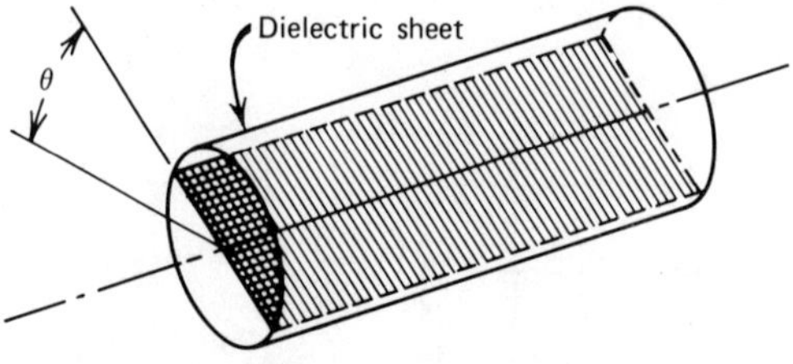

Figure 5.8. Schematic of half-wave plate.

Thus

$$b_1 = 0 \tag{5.27}$$

$$b_2 = 0 \tag{5.28}$$

$$b_3 = \frac{-e^{-j2\theta}}{\sqrt{2}} \tag{5.29}$$

$$b_4 = \frac{-je^{-j2\theta}}{\sqrt{2}} \tag{5.30}$$

The incoming and outgoing circularly polarized waves of the half-wave plate therefore differ by the phase angle 2θ, which is determined by the setting of the rotor position. It is also observed that the direction of rotation of the input wave has been reversed in traversing the half-wave plate.

Once the scattering matrices of the quarter- and half-wave plates are known, the operation of the variable rotor phase shifter in Figure 5.7 may be described by forming the input/output relation for each plate one at a time.

For the first section the input/output relation is defined by the quarter-wave plate in Chapter 4 by Eq. 4.52. The input/output relation of the half-wave plate described by Eq. 25 is next employed to construct the output waves of that section defined by Eqs. 5.27 through 5.30. These waves now form the input waves for the last quarter-wave plate. The output waves of the variable rotor phase shifter are therefore

$$\begin{bmatrix} b_1 \\ b_2 \\ b_3 \\ b_4 \end{bmatrix} = \begin{bmatrix} 0 & 0 & 1 & 0 \\ 0 & 0 & 0 & -j \\ 1 & 0 & 0 & 0 \\ 0 & -j & 0 & 0 \end{bmatrix} \begin{bmatrix} \dfrac{-e^{-j2\theta}}{\sqrt{2}} \\ \dfrac{-je^{-j2\theta}}{\sqrt{2}} \\ 0 \\ 0 \end{bmatrix} \tag{5.31 }$$

The result is

$$b_1 = 0 \tag{5.32}$$

$$b_2 = 0 \tag{5.33}$$

$$b_3 = \frac{-e^{-j2\theta}}{\sqrt{2}} \tag{5.34}$$

$$b_4 = \frac{-e^{-j2\theta}}{\sqrt{2}} \tag{5.35}$$

The two waves at ports 3 and 4 of the circular waveguide network re-combine at the rectangular waveguide as

$$-e^{-j2\theta} \tag{5.36}$$

where only the variable phase angle 2θ is retained for simplicity.

Thus the input wave is phase shifted through an angle 2θ that is simply determined by the angle the rotor section makes with respect to the two stator dielectric vanes.

5.5 *HIGH-POWER VARIABLE PHASE SHIFTER*

A 2-port variable phase shifter may be obtained by terminating the two output ports of a 3dB hybrid by a ganged shortcircuit piston. A commercial version is shown in Figure 5.9. The operation of this device can easily be understood by relating the incident and outgoing waves at the terminals of the hybrid through its scattering matrix.

Figure 5.9. Commercial version of high-power waveguide phase shifter (courtesy M.D.L.).

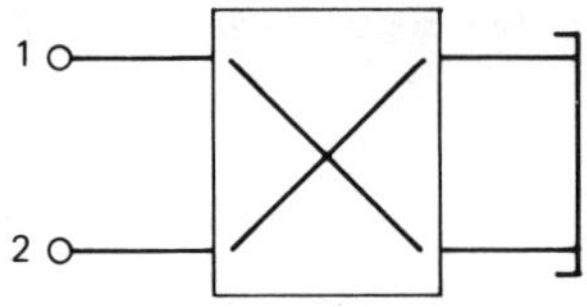

Figure 5.10. Schematic diagram of high-power phase shifter using 3-dB hybrid.

The incoming waves at the terminals of the device depicted in Figure 5.10 are

$$a_1 = 1 \tag{5.37}$$

$$a_2 = 0 \tag{5.38}$$

$$a_3 = b_3 e^{-j\theta} \tag{5.39}$$

$$a_4 = b_4 e^{-j\theta} \tag{5.40}$$

The outgoing waves of the device are given by

$$\begin{bmatrix} b_1 \\ b_2 \\ b_3 \\ b_4 \end{bmatrix} = \frac{1}{\sqrt{2}} \begin{bmatrix} 0 & 0 & 1 & j \\ 0 & 0 & j & 1 \\ 1 & j & 0 & 0 \\ j & 1 & 0 & 0 \end{bmatrix} \begin{bmatrix} 1 \\ 0 \\ b_3 e^{-j\theta} \\ b_4 e^{-j\theta} \end{bmatrix} \tag{5.41}$$

The result is

$$b_1 = \frac{1}{\sqrt{2}}(b_3 + jb_4)e^{-j\theta} \tag{5.42}$$

$$b_2 = \frac{1}{\sqrt{2}}(jb_3 + b_4)e^{-j\theta} \tag{5.43}$$

$$b_3 = \frac{1}{\sqrt{2}} \tag{5.44}$$

$$b_4 = \frac{j}{\sqrt{2}} \tag{5.45}$$

Eliminating b_3 and b_4 in the first two of these equations gives

$$b_1 = 0 \tag{5.46}$$

$$b_2 = je^{-j\theta} \tag{5.47}$$

Thus the outgoing wave at port 2 is related to the incoming one at port 1 by

$$b_2 = a_1 e^{-j\theta} \tag{5.48}$$

where θ is determined by the location of the shortcircuit piston.

PROBLEMS

1. Show that Eq. 5.25 satisfies the unitary condition.
2. Discuss the essential difference between the scattering matrix of the hybrid coupler in Chapter 3 and that of the quarter-wave plate in this chapter.
3. Satisfy $B=0$ in Eq. 5.5 in terms of A_i, B_i, C_i, and D_i.
4. Indicate why similar quarter-wave plates are used at each end of the rotary vane phase shifter instead of one of each kind described in Chapter 4.
5. Derive the input/output relation of the rotary vane phase shifter using the quarter-wave plate described in Chapter 4 by Eq. 4.58 instead of Eq. 4.52 used in this chapter.

CHAPTER SIX

Cavity Resonators

Cavity resonators find application as elements in bandpass and bandstop filter networks, as wavemeters and as resonant circuits in oscillators. The microwave cavity differs from its low-frequency version in that it relies on distributed lines or waveguides with magnetic or electric walls, instead of on lumped element *LC* resonators.

Since the main emphasis of this text is on the circuit aspects of microwave engineering the chapter starts by defining the unloaded, external and loaded quality factors of resonators, and introduces the notion of the immittance slope parameter of networks. It proceeds by developing the equivalence between lumped resonators and distributed open and short-circuited transmission lines. Field patterns of cavity resonators are introduced in the context of the rectangular waveguide cavity. The scattering matrix of the useful cavity using immittance discontinuities spaced half a wavelength apart is developed separately. Derivation of the immittance matrix for the symmetrical transformer coupled resonator completes the chapter.

6.1 *UNLOADED, EXTERNAL, AND LOADED Q-FACTORS*

It is usual in the study of low-frequency lumped element resonators to specify the losses in the circuit by means of a Q factor defined as

$$Q=\frac{\omega_0(\text{energy stored in the circuit})}{\text{power dissipated}} \tag{6.1}$$

In defining the Q factor the power that is dissipated within the circuit is normally expressed separately from that dissipated within the external circuit. For the simple LCr series circuit shown in Figure 6.1, the energy stored is

$$\tfrac{1}{2}LI^2 \tag{6.2}$$

whereas the power dissipated in the circuit at resonance is given by

$$\tfrac{1}{2}rI^2 \tag{6.3}$$

The unloaded Q factor Q_u is therefore given by

$$Q_u = \frac{\omega_0\left(\tfrac{1}{2}LI^2\right)}{\tfrac{1}{2}rI^2} \tag{6.4}$$

Thus

$$Q_u = \frac{\omega_0 L}{r} \tag{6.5}$$

If coupling is introduced to the resonator, the total power dissipated becomes

$$\tfrac{1}{2}I^2\left(r + n_1^2 Z_0 + n_2^2 Z_0\right) \tag{6.6}$$

where Z_0 is the load termination, and $n_{1,2}$ define the couplings to the resonators in Figures 6.2 and 6.3. Coupling to the distributed cavity resonators in Figure 6.4, take the form of either loops or probes. The loaded Q factor Q_L for the arrangements in Figures 6.2 and 6.3 is

$$Q_L = \frac{\omega_0\left(\tfrac{1}{2}LI^2\right)}{\tfrac{1}{2}I^2\left(r + n_1^2 Z_0 + n_2^2 Z_0\right)} \tag{6.7}$$

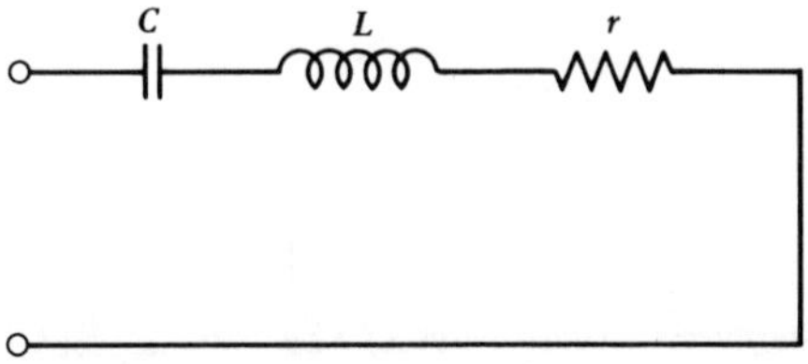

Figure 6.1. Schematic diagram of series LCr network.

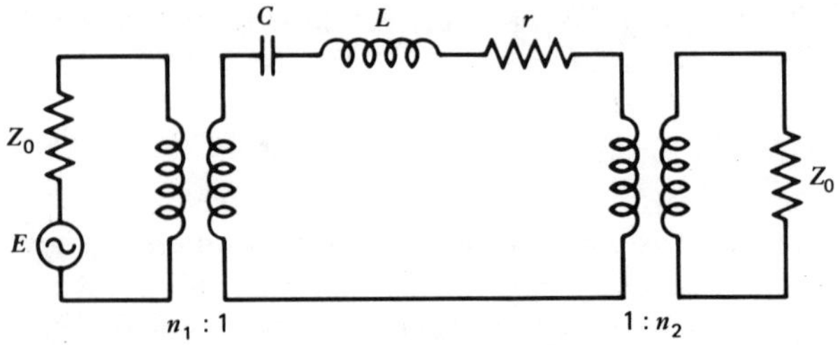

Figure 6.2. Schematic diagram of transformer coupled series LCr network.

Rewriting this equation gives

$$\frac{1}{Q_L} = \frac{1}{Q_u} + \frac{1}{Q_{e,1}} + \frac{1}{Q_{e,2}} \tag{6.8}$$

where

$$Q_{e1,2} = \frac{\omega_0\left(\frac{1}{2}LI^2\right)}{\frac{1}{2}n_{1,2}^2 Z_0 I^2} = \frac{\omega_0 L}{n_{1,2}^2 Z_0} \tag{6.9}$$

$Q_{e1,2}$ are the external Q factors of the circuit.

The Q factor defined by Eq. 6.1 can also be rewritten in a number of different ways. One particularly useful microwave definition for a series resonator is

$$Q_L = \frac{X'}{R} \tag{6.10}$$

where

$$X' = \frac{\omega_0}{2} \frac{dX}{d\omega}\bigg|_{\omega=\omega_0} \tag{6.11}$$

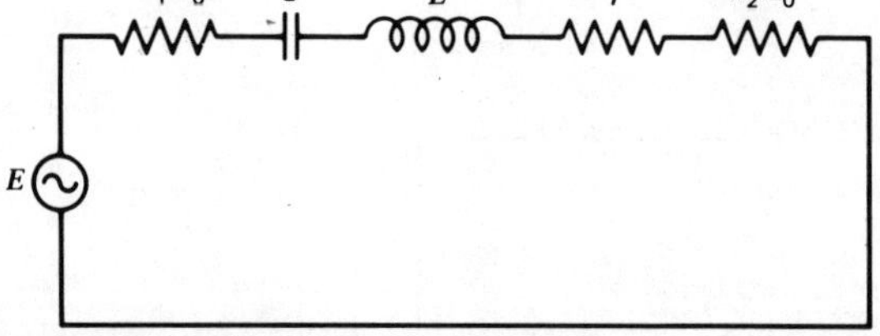

Figure 6.3. Equivalent circuit of transformer coupled series LCr network.

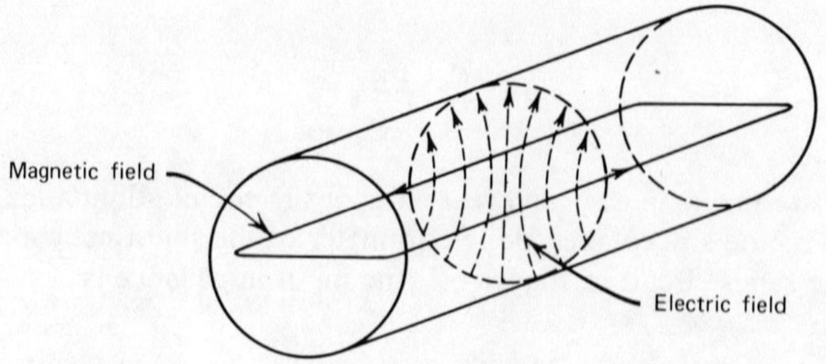

Figure 6.4a. Microwave cylindrical waveguide cavity resonator.

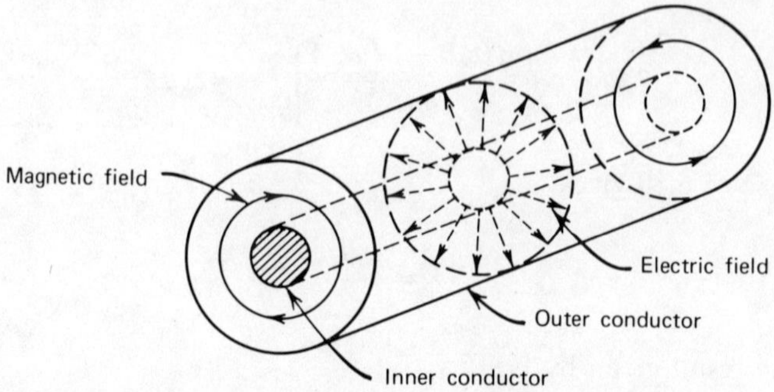

Figure 6.4b. Microwave coaxial resonator.

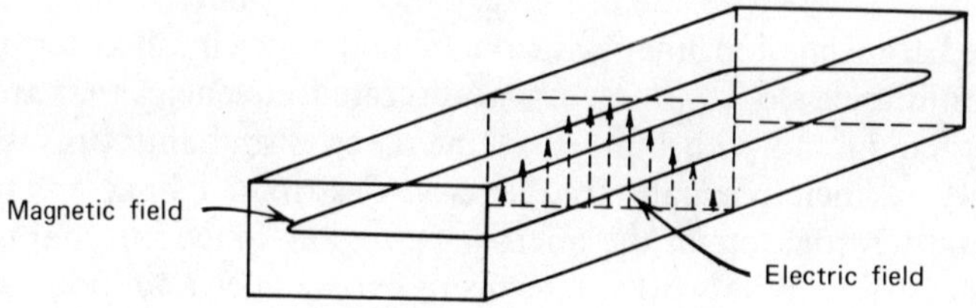

Figure 6.4c. Microwave rectangular waveguide cavity resonator.

X and R are the imaginary and real parts of the circuit impedance, and X' is known as the reactance slope parameter of the circuit.

A dual equation that applies to a shunt network is

$$Q_L = \frac{B'}{G} \tag{6.12}$$

where

$$B' = \frac{\omega_0}{2} \cdot \frac{dB}{d\omega}\bigg|_{\omega=\omega_0} \tag{6.13}$$

B and G are the imaginary and real parts of the circuit admittance, and B' is known as the susceptance slope parameter of the shunt network.

For the series circuit in Figure 6.1, the input impedance is

$$Z = r + j\omega L - \frac{j}{\omega C} \tag{6.14}$$

Using the definition for the reactance slope parameter in Eq. (6.11) gives

$$X' = \frac{\omega_0}{2} \frac{d(\omega L - 1/\omega C)}{d\omega}\bigg|_{\omega=\omega_0} = \omega_0 L \tag{6.15}$$

Making use of 6.10 gives

$$Q_u = \frac{\omega_0 L}{r} \tag{6.16}$$

This is the result given by Eq. 6.5.

6.2 *RESONANT CIRCUIT USING DISTRIBUTED TRANSMISSION LINES*

Distributed transmission lines with open- and shortcircuited terminations exhibit immittance slope parameters at discrete frequencies that are similar to those encountered with lumped element series and shunt resonators in the vicinity of their resonant frequencies. Distributed lines are therefore used as cavity resonators in the microwave region. Since it is inappropriate to discuss distributed networks in terms of capacitance and inductance, the equivalence between the two types of circuits is usually stated in terms of their immittance slope parameters. The two circuits are equivalent provided that their slope parameters are identical. This statement may be readily demonstrated by developing the equivalence between the impedance functions of the series lumped element and distributed opencircuited transmission line in Figures 6.5*a* and 6.5*b*.

The input impedance of the opencircuited distributed transmission line is

$$Z_{\text{in}} = Z_0 \coth(\alpha + j\beta)l \tag{6.17}$$

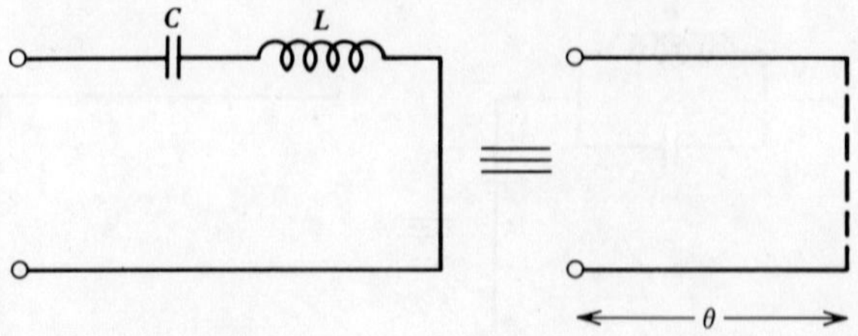

Figure 6.5. Equivalence between series LC circuit and opencircuited transmission line.

where α is the attenuation per unit length and β is the phase constant per unit length of the transmission line. Expanding the above equation in the vicinity of $\beta_0 l = \pi/2$ gives

$$Z_{in} \approx Z_0 \alpha l + \frac{jZ_0 \pi}{4}\left(\frac{\omega}{\omega_0} - \frac{\omega_0}{\omega}\right) \tag{6.18}$$

In obtaining this equation use is made of the fact that

$$\beta = \frac{\omega}{v} = \frac{\omega_0}{v} \cdot \frac{\omega}{\omega_0} = \beta_0\left[1 + \left(\frac{\omega - \omega_0}{\omega_0}\right)\right]$$

$$\approx \beta_0\left[1 + \frac{1}{2}\left(\frac{\omega}{\omega_0} - \frac{\omega_0}{\omega}\right)\right] \tag{6.19}$$

Equation 18 may be compared with that obtained by forming the input impedance of the series resonant circuit in Figure 6.1.

$$Z_{in} = r + j\omega_0 L\left(\frac{\omega}{\omega_0} - \frac{\omega_0}{\omega}\right) \tag{6.20}$$

It is readily observed that both impedance functions have the same nature in the vicinity of $\omega = \omega_0$. The equivalence between them is satisfied by making their reactance slope parameters equal.

Using the definition for the reactance slope parameter given by Eq. 6.11 leads to

$$\omega_0 L = \frac{\pi Z_0}{4} \tag{6.21}$$

The duality between the shunt resonator and distributed shortcircuited transmission line in Figure 6.6 follows in a similar fashion.

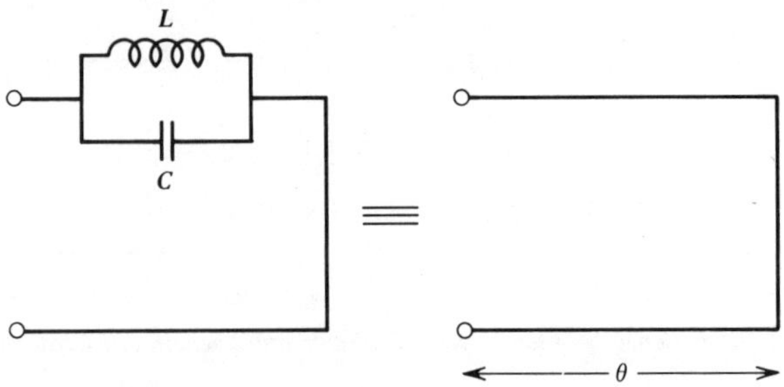

Figure 6.6. Equivalence between shunt LC network and shortcircuited transmission line.

6.3 BOUNDARY CONDITIONS OF RECTANGULAR MICROWAVE CAVITY RESONATORS

One model of the microwave cavity resonator is obtained by terminating a section of waveguide by shortcircuit planes. The boundary conditions for such a cavity may be obtained by starting with the reflection coefficient of a waveguide terminated by a discontinuity Γ

$$S'_{11} = S_{11} + \frac{S_{21}^2\Gamma}{1 - S_{22}\Gamma} \tag{6.22}$$

Introducing electric walls at the input and output terminals of the waveguide gives

$$\Gamma = -1 \tag{6.23}$$

$$S'_{11} = -1 \tag{6.24}$$

The boundary conditions of the cavity resonator are therefore given by

$$-1 = S_{11} - \frac{S_{21}^2}{1 + S_{22}} \tag{6.25}$$

For a 2-port transmission line,

$$S_{11} = S_{22} = 0 \tag{6.26}$$

$$S_{12} = S_{21} = \exp(-j\theta) \tag{6.27}$$

where θ is related to wavelength in the waveguide and the length of the

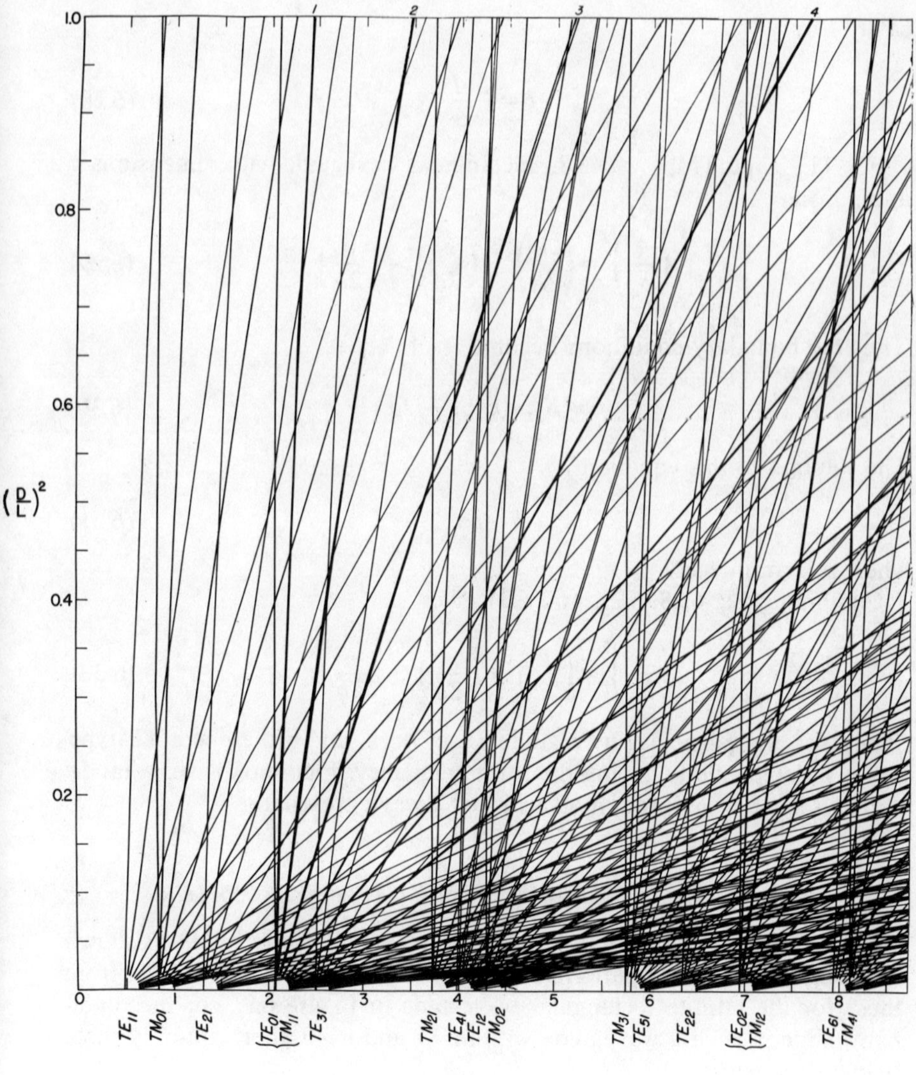

Figure 6.7. Mode chart for rectangular waveguide resonator.

cavity l by

$$\theta = \frac{2\pi l}{\lambda_g} \tag{6.28}$$

For TE_{mn} and TM_{mn} waves in rectangular waveguide with dimensions a and b, λ_g is

$$\left(\frac{1}{\lambda_g}\right)^2 = \left(\frac{1}{\lambda_0}\right)^2 - \left(\frac{m}{2a}\right)^2 - \left(\frac{n}{2b}\right)^2 \tag{6.29}$$

Thus the boundary conditions become

$$1 = S_{21}^2 = \exp(-j2\theta) \tag{6.30}$$

The solution to this equation is

$$\theta = p\pi \tag{6.31}$$

where p is an integer.

Combining Eq. 6.28, 6.29, and 6.31 gives

$$\left(\frac{1}{\lambda_0}\right)^2 = \left(\frac{P}{2l}\right)^2 + \left(\frac{m}{2a}\right)^2 + \left(\frac{n}{2b}\right)^2 \tag{6.32}$$

Thus, for any particular waveguide set of m, n, and p there is a TE-type and a TM type mode having the same frequency. Figure 6.7 gives a mode chart for some lower order modes in such a cavity resonator.

6.4 *RECTANGULAR WAVEGUIDE CAVITY RESONATOR*

The simplest type of cavity resonator is in rectangular waveguide, as shown in Figure 6.8. The field patterns for this cavity are readily determined from those for the infinite rectangular waveguide in Figure 6.9. For the dominant mode in such a waveguide with $m=1$ and $n=0$, there are only three field components:

$$H_z = \cos\frac{\pi x}{a} \cdot \exp\left(-\frac{j2\pi z}{\lambda_g}\right) \tag{6.33}$$

$$H_x = j\left(\frac{\lambda_c}{\lambda_g}\right)\sin\frac{\pi x}{a} \cdot \exp\left(-\frac{j2\pi z}{\lambda_g}\right) \tag{6.34}$$

$$E_y = -j\left(\frac{\lambda_c}{\lambda_0}\right)\sqrt{\frac{\mu_0}{\epsilon_0}}\,\sin\frac{\pi x}{a} \cdot \exp\left(-\frac{j2\pi z}{\lambda_g}\right) \tag{6.35}$$

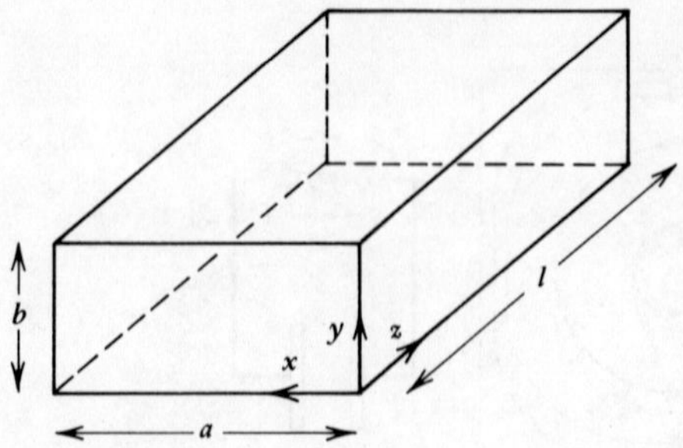

Figure 6.8. Rectangular waveguide resonator.

A common factor $\exp(-j\omega t)$ is assumed throughout.

Taking a linear combination of these field patterns gives the cavity modes as

$$H_z = j2\cos\frac{\pi x}{a}\sin\frac{\pi z}{l}\cdot\exp -j\left(\omega t+\frac{\pi}{2}\right) \tag{6.36}$$

$$H_x = j2\left(\frac{\lambda_c}{\lambda_g}\right)\sin\frac{\pi x}{a}\cos\frac{\pi z}{l}\cdot\exp -j\left(\omega t+\frac{\pi}{2}\right) \tag{6.37}$$

$$E_y = 2\left(\frac{\lambda_c}{\lambda_0}\right)\sqrt{\frac{\mu_0}{\epsilon_0}}\ \sin\frac{\pi x}{a}\sin\frac{\pi z}{l}\cdot\exp -j(\omega t) \tag{6.38}$$

Figure 6.10 indicates the field patterns in such a cavity at time intervals $\omega t = 0$, $\pi/2$, π, and $3\pi/2$, where the fields are purely electric, purely magnetic, and purely electric and magnetic in the opposite direction. The reason for this behavior is that in a perfectly conducting enclosure the

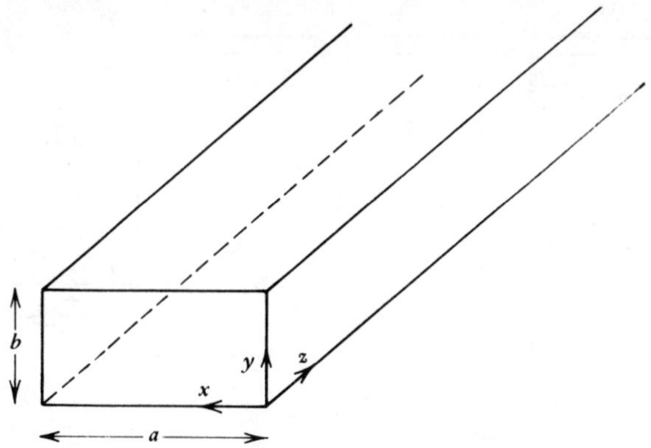

Figure 6.9. Section of rectangular waveguide.

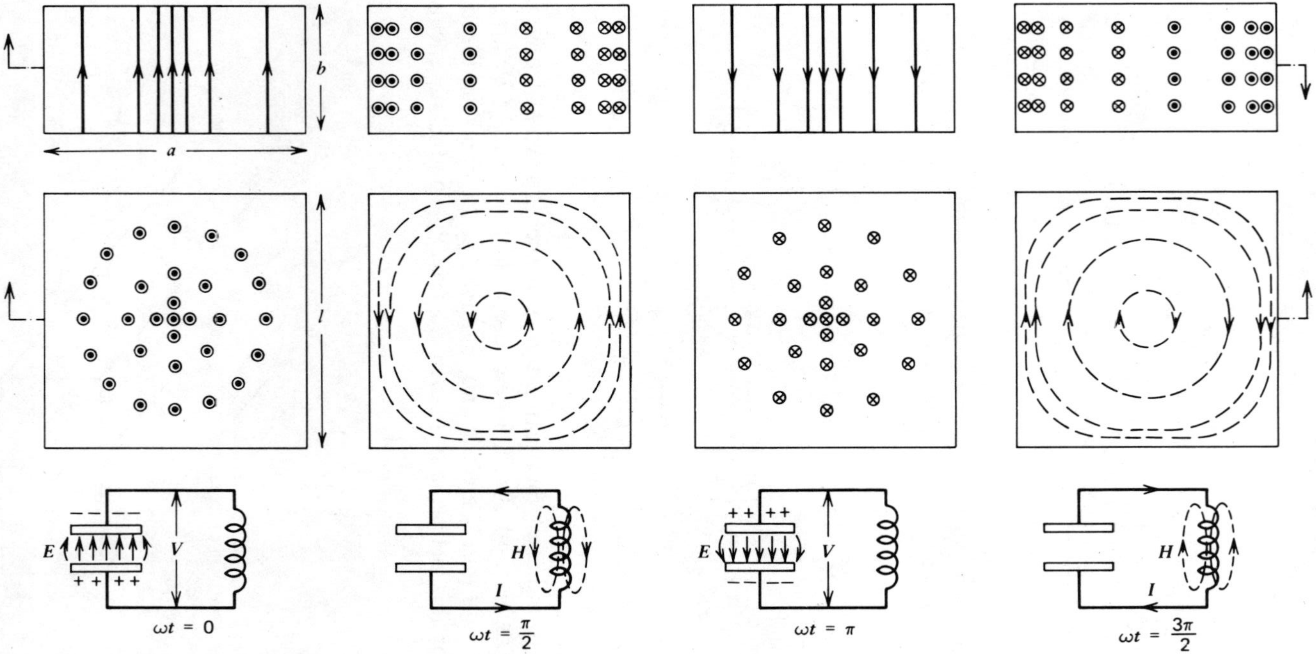

Figure 6.10. *Elevation and plan view of **E** and **H** fields in rectangular waveguide cavity and the corresponding voltage and current in LC circuit.*

electric and magnetic fields are in phase quadrature. This implies that there is a continuous exchange of energy between the electric and magnetic fields. Such an exchange of energy is also found in a simple lumped element resonator.

6.5 *SCATTERING MATRIX OF SERIES RESONATOR*

This section derives the scattering matrix of the series resonator indicated in Figure 6.11*a* using the method developed in Chapter 1. The matrix of this resonator is readily obtained by forming its two eigennetworks. The

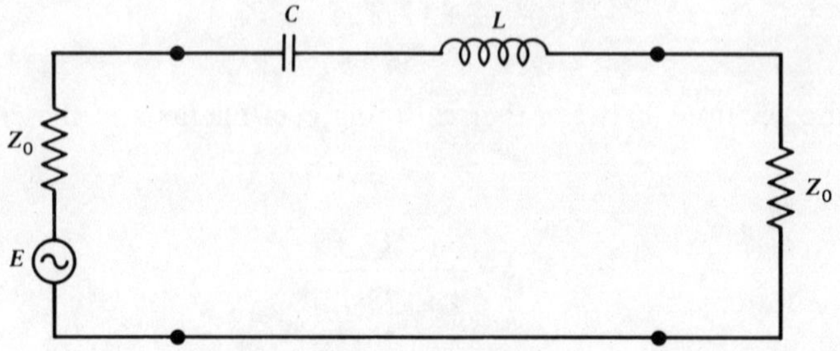

Figure 6.11a. Doubly loaded series LC network.

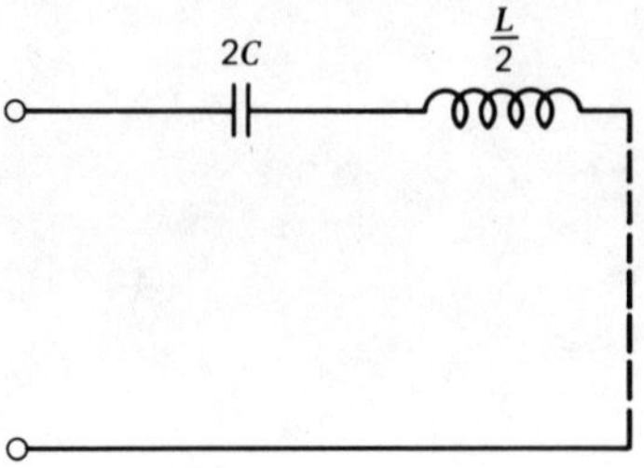

Figure 6.11b. In-phase eigennetwork of series LC network.

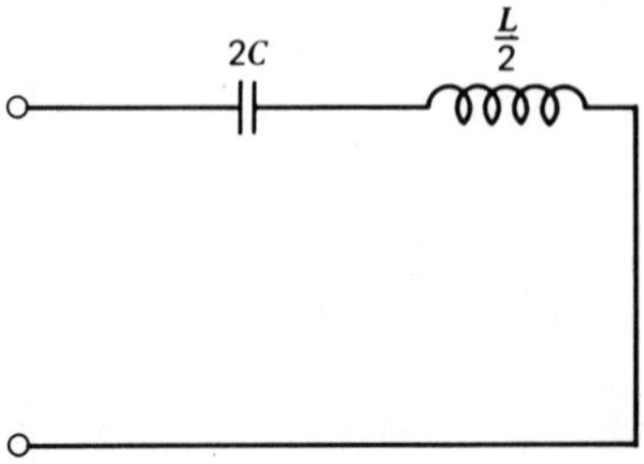

Figure 6.11c. Out of phase eigennetwork of LC network.

impedances of the two circuits in Figure 6.11*b* and 6.11*c* are

$$Z_1 = Z_{0/c} = \infty \tag{6.39}$$

$$Z_2 = Z_{s/c} = \frac{j\omega L}{2} + \frac{1}{j2\omega C} \tag{6.40}$$

Thus, the scattering eigenvalues are

$$s_1 = 1 \tag{6.41}$$

$$s_2 = \frac{(j\omega L/2 + 1/j2\omega C) - Z_0}{(j\omega L/2 + 1/j2\omega C) + Z_0} \tag{6.42}$$

Using the relations between the scattering coefficients and the eigenvalues gives:

$$|S_{11}| = \frac{(2\delta Q_L)^2}{1 + (2\delta Q_L)^2} \tag{6.43}$$

$$|S_{21}|^2 = \frac{1}{1 + (2\delta Q_L)^2} \tag{6.44}$$

where Q_L is the doubly terminated Q factor

$$Q_L = \frac{\omega_0 L}{2Z_0} \tag{6.45}$$

and

$$2\delta = \frac{\omega}{\omega_0} - \frac{\omega_0}{\omega} \tag{6.46}$$

$$\omega_0^2 LC = 1 \tag{6.47}$$

The Q factor determines the frequency on either side of the center frequency at which the transmitted power is half that at the main resonance. This is illustrated in Figure 6.12.

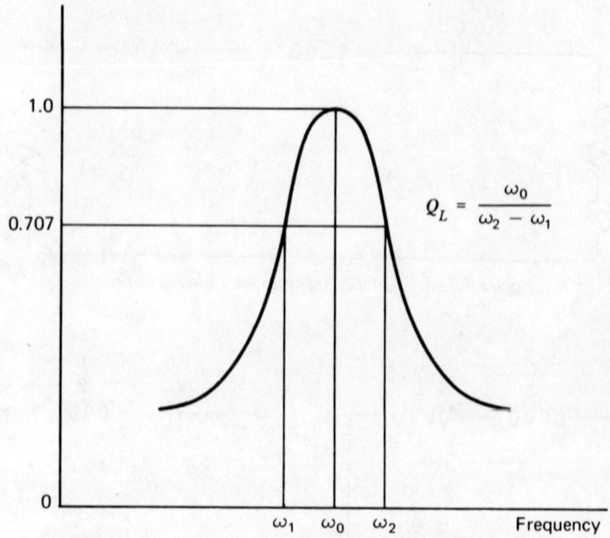

Figure 6.12. Frequency characteristic of LC resonator showing definition of loaded Q factor.

6.6 SCATTERING MATRIX OF SERIES RESONATOR WITH DAMPING

The development of the scattering matrix of the series resonator with damping shown in Figure 6.13*a* proceeds once more by forming its eigennetworks. In this instance the scattering coefficient S_{21} is not unity at resonance, but is related to the damping term as expected.

Constructing the impedance eigenvalues for the eigennetworks in Figures 6.13*b* and 6.13*c* gives

$$Z_1 = Z_{o/c} = \infty \tag{6.48}$$

$$Z_2 = Z_{s/c} = \frac{r}{2} + \frac{j\omega L}{2} + \frac{1}{j2\omega C} \tag{6.49}$$

Thus, the scattering eigenvalues are

$$s_1 = 1 \tag{6.50}$$

$$s_2 = \frac{(r/2 + j\omega L/2 + 1/j2\omega C) - Z_0}{(r/2 + j\omega L/2 + 1/j2\omega C) + Z_0} \tag{6.51}$$

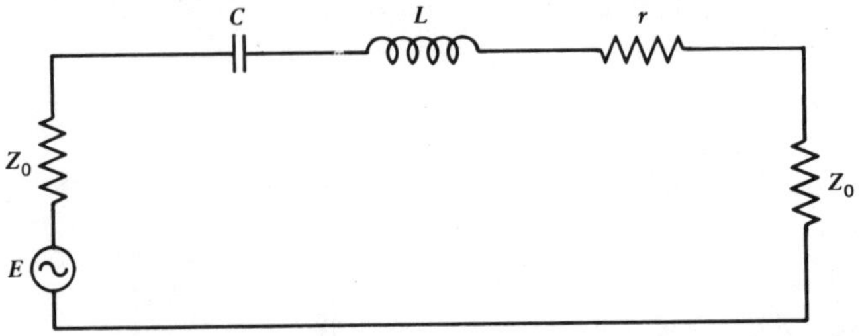

Figure 6.13a. Doubly terminated LCr network.

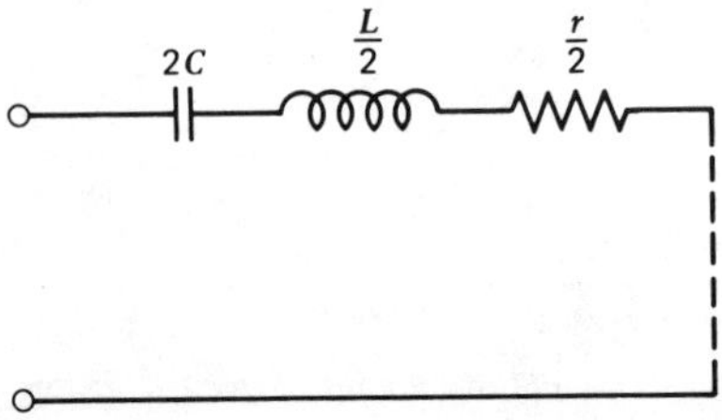

Figure 6.13b. In-phase eigennetwork of LCr network.

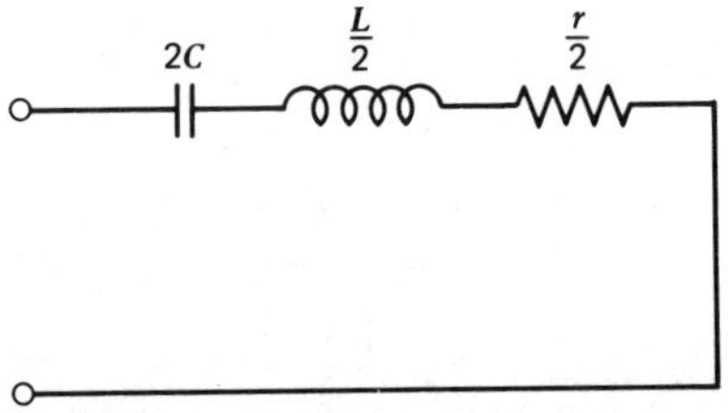

Figure 6.13c. Out-of-phase eigennetwork of LCr network.

S_{21} is therefore given by

$$S_{21}=\frac{s_1-s_2}{2}=\frac{2Z_0/r}{1+2Z_0/r+j\omega_0L/r[(\omega/\omega_0)-(\omega_0/\omega)]} \tag{6.52}$$

This equation is also sometimes written in terms of an external coupling coefficient β_e

$$S_{21}=\frac{2\beta_e}{1+2\beta_e+jQ_u[(\omega/\omega_0)-(\omega_0/\omega)]} \tag{6.53}$$

where β_e is

$$\beta_e=\frac{Z_0}{r}=\frac{Q_u}{Q_e} \tag{6.54}$$

Q_u is defined by Eq. 6.5 and Q_e by Eq. 6.9 with $n_{1,2}=1$.

An alternate form for S_{21} is

$$S_{21}=\frac{S_{21}(\omega_0)}{1+jQ_L[(\omega/\omega_0)-(\omega_0/\omega)]} \tag{6.55}$$

where

$$S_{21}(\omega_0)=\frac{2\beta_e}{1+2\beta_e} \tag{6.56}$$

and Q_L is defined by Eq. 6.8 with $n_{1,2}=1$. The factor 2 in the above equation is omitted when Q_e is defined as the doubly loaded Q factor. The derivation of S_{11} is left as an exercise for the reader.

6.7 *MICROWAVE CAVITY RESONATOR USING SUSCEPTANCES SPACED BY SECTION OF LINE*

One standard microwave cavity resonator employed in the construction of high-quality bandpass filters consists of two susceptances spaced by a section of transmission line in the manner depicted in Figure 6.14*a*. The resonant frequency of this cavity is fixed by adjusting the spacing between the two susceptances, whereas its loaded Q factor is determined by the value of the susceptances. The derivation of these two quantities for a symmetrical cavity proceeds by using the eigennetwork approach.

The eigenadmittance for the in-phase circuit in Figure 6.14*b* is

$$Y_1=j\left(B+Y_0\tan\frac{\theta}{2}\right) \tag{6.57}$$

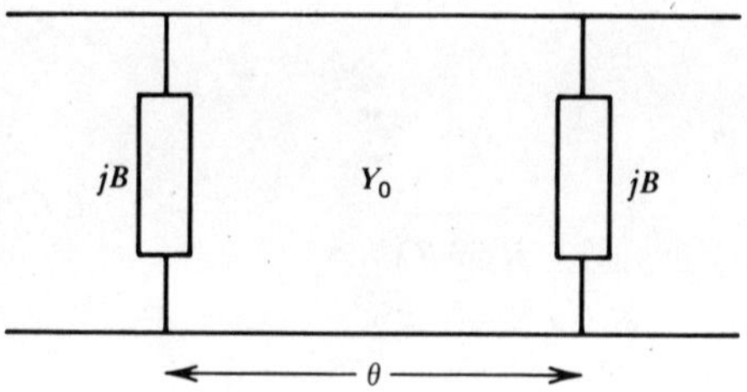

Figure 6.14a. Schematic diagram of cavity resonator using shunt susceptances spaced by section of transmission line.

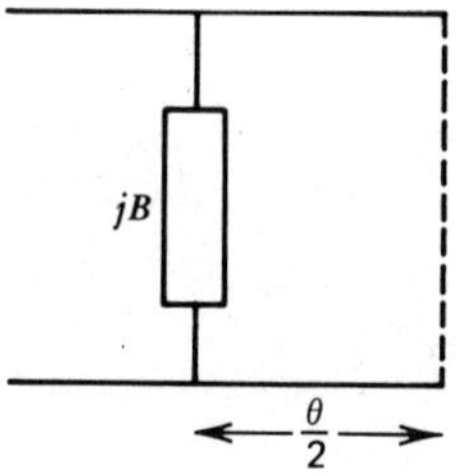

Figure 6.14b. In-phase eigennetwork of cavity resonator formed by susceptances spaced by section of transmission line.

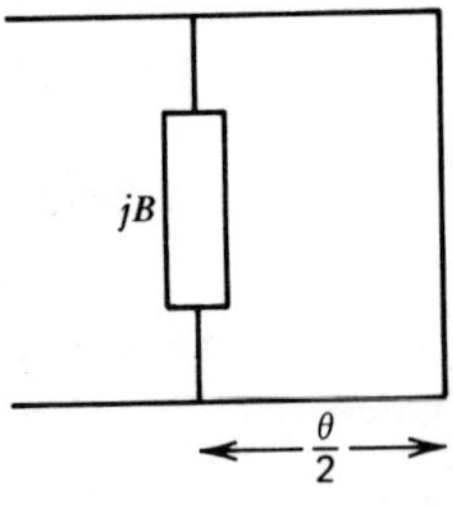

Figure 6.14c. Out-of-phase eigennetwork of cavity resonator formed by susceptances spaced by section of transmission line.

For the out-of-phase eigennetwork in Figure 6.14c, the eigenadmittance is

$$Y_2 = j\left(B - Y_0 \cot\frac{\theta}{2}\right) \tag{6.58}$$

The reflection coefficients for the two eigennetworks are given in the usual way by

$$s_1 = \frac{Y_0 - Y_1}{Y_0 + Y_1} \tag{6.59}$$

$$s_2 = \frac{Y_0 - Y_2}{Y_0 + Y_2} \tag{6.60}$$

Writing S_{21} in terms of s_1 and s_2 gives

$$S_{21} = \frac{s_1 - s_2}{2} \tag{6.61}$$

Thus the final result is

$$S_{21}^2 = \frac{1}{1 + (B^2/4)(2\cos\theta - B\sin\theta)^2} \tag{6.62}$$

The resonant frequency of the cavity is defined by $S_{21} = 1$.

$$\tan\theta_0 = \frac{2}{B} \tag{6.63}$$

The relation between the susceptance B and the loaded Q factor Q_L of the overall cavity may now be determined by putting S_{21} in the standard form described by Eq. 6.44. This may be done by expanding θ in Eq. 6.62 in a Taylor series in the vicinity of the resonant frequency defined by Eq. 6.63.

Expanding $\sin\theta$ and $\cos\theta$ in a Taylor series gives

$$\sin(\theta_0+\Delta\theta)\simeq\sin\theta_0+\Delta\theta\cos\theta_0 \tag{6.64}$$

$$\cos(\theta_0+\Delta\theta)\simeq\cos\theta_0-\Delta\theta\sin\theta_0 \tag{6.65}$$

At resonance Eq. 6.63 gives

$$\sin\theta_0=\frac{2}{\sqrt{B^2+4}} \tag{6.66}$$

$$\cos\theta_0=\frac{B}{\sqrt{B^2+4}} \tag{6.67}$$

Thus, near $\tan\theta_0=2/B$, S_{21} becomes

$$S_{21}^2=\frac{1}{1+(\pi^2B^2/16)(B^2+4)(2\delta)^2} \tag{6.68}$$

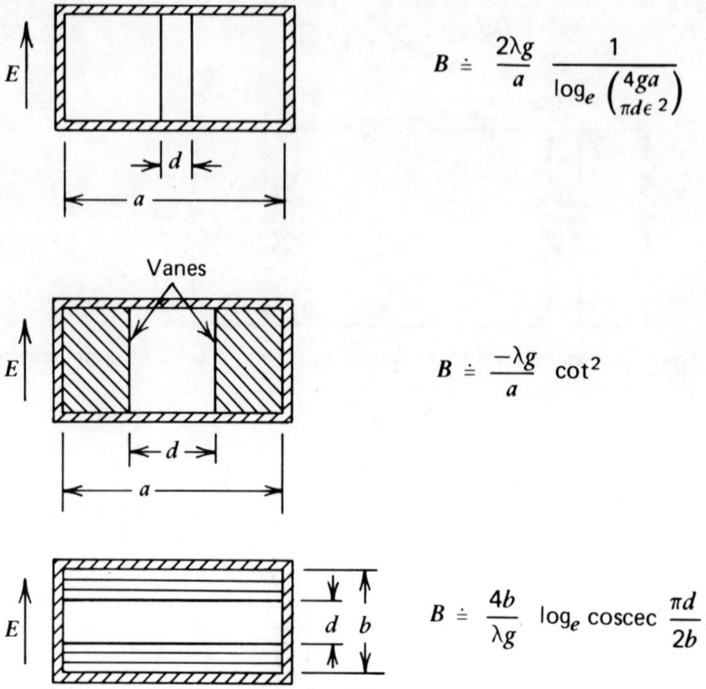

Figure 6.15. Microwave immittance structures.

provided θ_0 is expanded about π instead of $\tan\theta_0 = 2/B$.

Comparing the above equation with the standard form defined by Eq. 6.44 gives the desired result

$$Q_L = \frac{\pi B}{4}\sqrt{(B^2+4)} \tag{6.69}$$

Figure 6.15 depicts three practical types of waveguide susceptances.

6.8 *IMPEDANCE MATRIX OF TRANSFORMER-COUPLED RESONATOR*

The Z matrix of the symmetrical transformer-coupled lumped resonator in Figure 6.16*a* is also readily obtained from its eigennetworks shown in Figures 6.16*b* and 6.16*c*. For the in-phase eigennetwork the impedance eigenvalue is

$$Z_1 = j\omega L_1 \tag{6.70}$$

The impedance eigenvalue for the out-of-phase eigennetwork is

$$Z_2 = j\omega L_1 - \frac{2\omega^2 M^2}{j\omega L + 1/j\omega C} \tag{6.71}$$

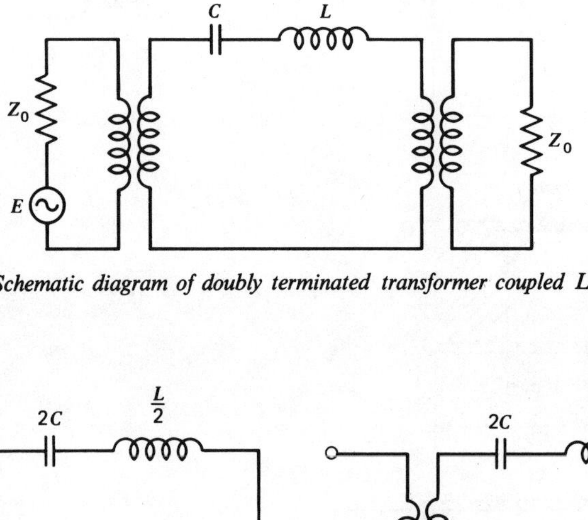

Figure 6.16a. Schematic diagram of doubly terminated transformer coupled LC series resonator.

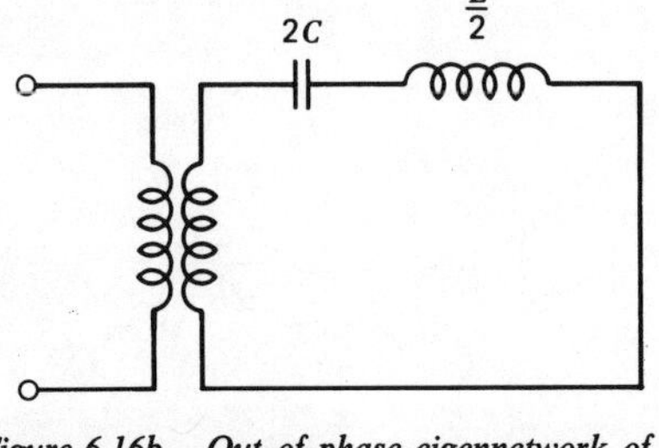

Figure 6.16b. Out-of-phase eigennetwork of transformer coupled series LC resonator.

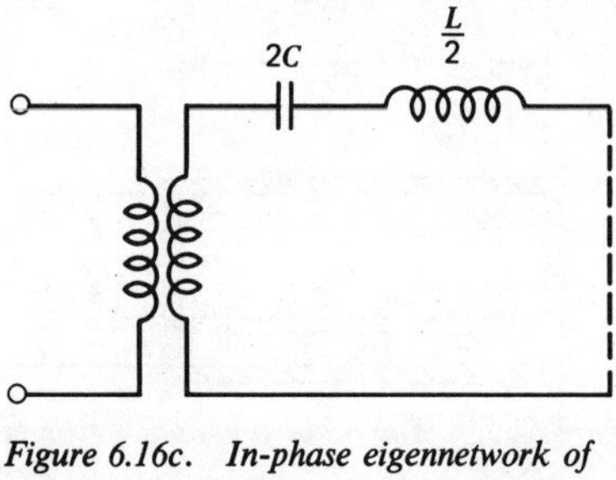

Figure 6.16c. In-phase eigennetwork of transformer coupled series LC resonator.

Thus

$$Z_{11}=Z_{22}=\frac{Z_1+Z_2}{2}=j\omega L_1+\frac{j\omega^3M^2/L}{-\omega^2+\omega_0^2} \tag{6.72}$$

$$Z_{21}=Z_{12}=\frac{Z_1-Z_2}{2}=\frac{-j\omega^3M^2/L}{-\omega^2+\omega_0^2} \tag{6.73}$$

The equivalent T circuit for this network is left as an exercise for the reader.

Introducing damping in the above equations gives

$$Z_{11}=Z_{22}=j\omega L_1+\frac{j\omega^3M^2/L}{-\omega^2+\omega_0^2+j\omega_0\omega/Q_u} \tag{6.74}$$

$$Z_{21}=Z_{12}=\frac{-j\omega^3M^2/L}{-\omega^2+\omega_0^2+j\omega_0\omega/Q_u} \tag{6.75}$$

where Q_u is the unloaded Q factor of the resonator in 6.5 or 6.16

PROBLEMS

1. Show that $Q_L=\omega_0/\Delta\omega$, where $\Delta\omega$ is the width of the resonance curve at which the transmitted power is half that at resonance.
2. Obtain the scattering matrix for a shunt LC resonator.
3. Show that the reactance slope parameter of a quarter-wave-long open-circuited transmission line having an input impedance $jZ_0\cos\beta l$ is given by $\pi Z_0/4$.
4. Demonstrate that the input impedance of a lumped element series resonator may be written in terms of stored electric and magnetic energies and power dissipated as

$$Z=\frac{[P_{\text{diss}}+j2\omega(W_e-W_m)]}{|I_0|^2}$$

5. Show the equivalence between a lumped element shunt resonator and a quarter-wave-long shortcircuited transmission line.
6. Obtain the susceptance slope parameter of a quarter-wave-long short-circuited transmission line for which the input admittance is $jY_0\tan\beta l$.

7. Use the equivalent circuit in Figure 6.3 to obtain S_{21} in the form given by Eq. 6.55.
8. Obtain S_{11} for the cavity in Figure 6.14 with the help of Eq. 6.62 and the unitary condition.
9. Derive the scattering matrix of a shunt LC resonator with damping.
10. Obtain the equivalent T circuit for the transformer-coupled resonator described by Eqs. 6.72 and 6.73.

FURTHER READING

1. R. F. Harrington, *Time-Harmonic Electromagnetic Fields*, McGraw-Hill, New York, 1961.

CHAPTER 7

Microwave Filters

Microwave filters form one of the most important classes of circuits used in microwave engineering. Depending on system requirement they may have lowpass, bandpass, bandstop, or highpass frequency characteristics. The modern filter theory, developed by Darlington,[1] relies on synthesis, whereby the nature of the transfer function is specified and the corresponding network is determined as a reactance network terminated in a 1Ω resistor.

The first problem in network synthesis is to determine what properties S_{21} must have to represent a network composed of L's and C's. It can be shown that the impedance function associated with the transfer function must be positive real. The second part of network synthesis is to find methods whereby the corresponding network can be found. This section is mainly concerned with the latter problem for filters with Butterworth (or maximally flat specifications). The conventional approach to filter theory is to develop a lowpass prototype ladder network normalized to a 1-ohm termination and a cut-off frequency of 1 rad/sec. Frequency and impedance transformations are then used to derive highpass, bandpass, and bandstop filters. This avoids the need to set down a multiplicity of tabulated results for the many different filter specifications met in practice. Since the final ladder networks are lumped LC networks, their microwave realizations involve a statement of loaded Q factors and microwave immittances instead of lumped L's and C's.

7.1 *THE SYNTHESIS PROCEDURE*

If the transfer function is stated in terms of the scattering parameters the first task in the synthesis problem is to obtain the input driving immittance

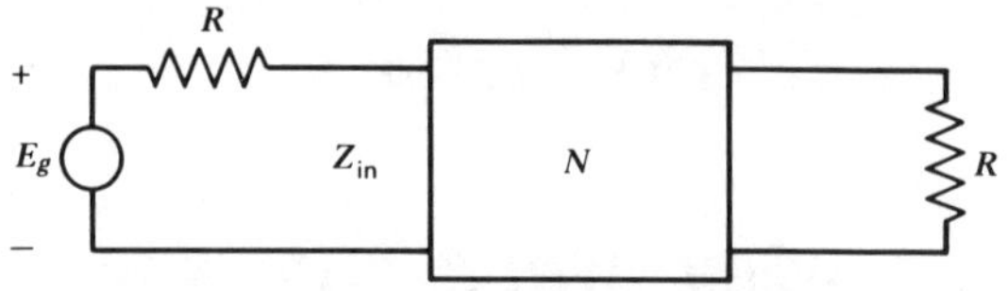

Figure 7.1. Doubly terminated filter circuit.

of the network from the nature of the transfer function. For the lossless 2-port network between 1 Ω terminations shown in Figure 7.1, S_{11} and S_{21} are related by

$$|S_{11}(j\omega)|^2 = 1 - |S_{21}(j\omega)|^2 \tag{7.1}$$

where ω is the normal frequency variable.

Introducing the complex frequency variable $s = j\omega$ and making use of analytical continuity gives

$$S_{11}(s)S_{11}(-s)|_{s=j\omega} = 1 - |S_{21}(j\omega)|^2 \tag{7.2}$$

In what follows the required solution is associated with $S_{11}(s)$ since the poles of the immittance function must lie on the left half of the s plane for the immittance function to be p.r.

Once $S_{11}(s)$ is known the input immittance of the network is given by the following standard relation

$$Y_{in}(s) = \frac{1 - S_{11}(s)}{1 + S_{11}(s)} \tag{7.3}$$

The final step involves constructing $Y_{in}(s)$ as a ladder network. Since transformations based on a lowpass prototype are conventionally used to obtain the highpass, bandpass, and bandstop filters it is sufficient to construct the lowpass prototype. A canonical realization for the lowpass prototype is usually obtained by forming a Cauer-type ladder expansion of $Y_{in}(s)$ or $Z_{in}(s)$

7.2 *BUTTERWORTH LOWPASS FILTER APPROXIMATION*

Since an ideal lowpass filter characteristic is not possible some type of approximation to it is necessary. The two most often used approximations are the Butterworth and Chebyshev ones. For simplicity only the Butterworth one will be considered. This transfer function is defined by

$$|S_{21}(j\omega)|^2 = \frac{1}{1+\omega^{2n}} \tag{7.4}$$

This function has a 3-dB point at $\omega = 1$ for all n, and its amplitude response falls off at a rate of $6n$-dB/octave. It also has the characteristic that the first $2n-1$ derivatives are equal to zero at the origin $\omega = 0$. This amplitude response is illustrated in Figure 7.2. The Chebyshev transfer function is defined by Eq. 4.15. The poles of this function in the complex plane are given by

$$1+(-s^2)^n = 0 \tag{7.5}$$

Therefore, the poles location are given by

$$s_k = (-1)^{1/2}(-1)^{n/2} = \exp\left[j\frac{2k+n-1}{n}\cdot\frac{\pi}{2}\right] \tag{7.6}$$

where $k = 1, 2, 3, \ldots, 2n$.

These poles lie on a unit circle in the s plane and are symmetrical with both the real and imaginary axes. For a positive real immittance the poles of the adopted solution lie on the left half of the s plane.

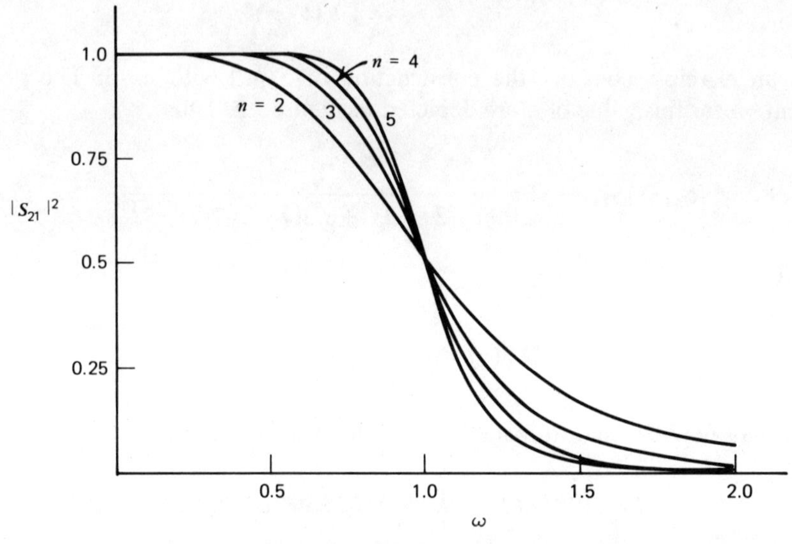

Figure 7.2. Lowpass Butterworth transfer characteristic.

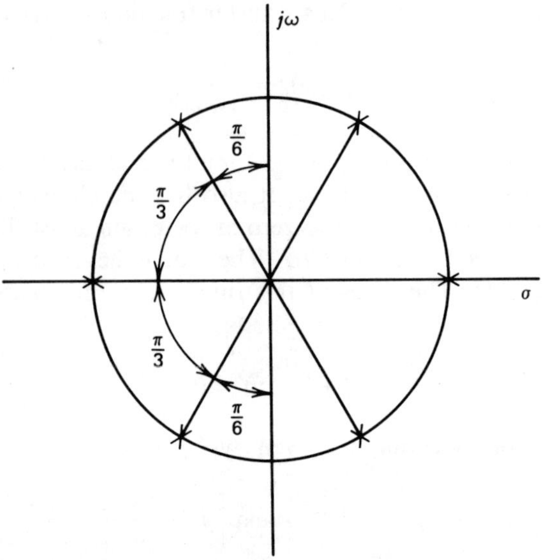

Figure 7.3. Poles of S_{11} (s) and S_{11} (− s) for n=3 Butterworth filter.

Writing $S_{11}(s)$ in terms of $S_{21}(j\omega)$ by replacing s by $j\omega$ gives

$$S_{11}(s)S_{11}(-s)=\frac{(-s^2)^n}{1+(-s^2)^n} \tag{7.7}$$

As an example consider the construction of $S_{11}(s)$ with $n=3$. The pole location for this value of n are depicted in Figure 7.3. Thus

$$S_{11}(s)S_{11}(-s)=\frac{s^3(-s^3)}{(1+2s+2s^2+s^3)(1-2s+2s^2-s^3)} \tag{7.8}$$

and

$$S_{11}(s)=\frac{\pm s^3}{1+2s+2s^2+s^3} \tag{7.9}$$

S_{11} may now be used to form Y_{in} with the help of Eq. 7.3.

7.3 *DARLINGTON INSERTION LOSS FILTER SYNTHESIS*

Once $S_{11}(s)$ has been determined, the input admittance of the network can be formed by combining Eq. 7.3 and 7.9. Using the minus sign in Eq. 7.9

gives

$$Y_{\text{in}}(s) = \frac{2s^3 + 2s^2 + 2s + 1}{2s^2 + 2s + 1} \tag{7.10}$$

A canonical realization for $Y_{\text{in}}(s)$ may now be obtained by performing a Cauer type ladder expansion of the admittance function

$$2s^2 + 2s + 1 \overline{)2s^3 + 2s^2 + 2s + 1} \; (s \rightarrow y$$

$$\underline{2s^3 + 2s^2 + s}$$

$$s + 1)\, 2s^2 + 2s + 1 (2s \rightarrow z$$

$$\underline{2s^2 + 2s}$$

$$1)s + 1 (s \rightarrow y$$

$$\underline{s}$$

$$1)1(1$$

The low pass filter is thus synthesized in the structure indicated in Figure 7.4. This filter structure is normalized to a cut-off frequency of 1 rad/sec and to 1 Ω terminations.

The synthesis of filter networks with different values of n proceeds in a similar way, but one recurrence formula for the elements of the filter is given by

$$g_r = 2\sin\left[\frac{(2r-1)\pi}{2n}\right] \qquad r = 1, 2, \ldots, n \tag{7.11}$$

For $n=3$ this last equation gives $g_1 = 1, g_2 = 2, g_3 = 1$, which is in agreement with the result obtained above. Using the positive sign in Eq. 7.9 gives an equivalent lowpass T instead of Π circuit.

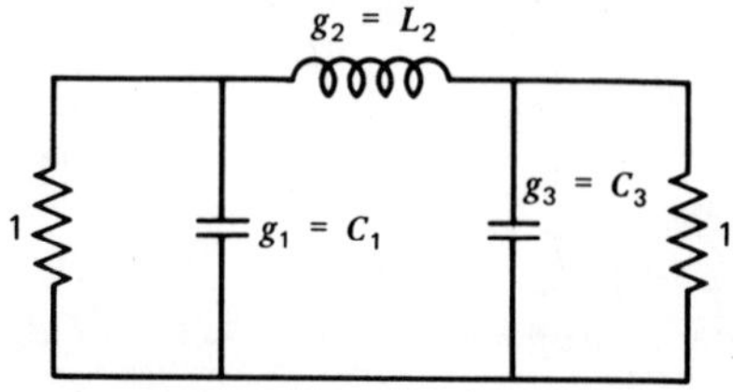

Figure 7.4. Lowpass prototype, $n=3$.

7.4 FREQUENCY TRANSFORMATIONS

Highpass, bandpass, and bandstop filters can be obtained from the lowpass prototype developed in the last section by a technique known as frequency transformation. Using frequency transformations, the elements of the normalized lowpass prototype are changed into elements of the highpass, bandpass, and bandstop, which includes denormalization of the cut-off frequency so that the filters need only be scaled for impedance.

The lowpass to highpass transformation is obtained by replacing s' by ω_0/s.

$$s' \to \frac{\omega_0}{s} \tag{7.12}$$

where s' represents the normalized lowpass frequency variable, s is the regular frequency variable, and ω_0 the normalizing constant is dimensionless and is often taken to be the actual cut-off frequency of the highpass filter. This transformation maps the segment $0<|\omega'|<1$ on the s' plane to $\omega_0<|\omega|<\infty$ on the s plane in the manner indicated in Figure 7.5.

The relation between the lowpass and highpass networks is obtained by noting that immittance is invariable under frequency transformation. For the inductance the impedance is

$$z = s'L \tag{7.13}$$

Introducing the frequency transformation in the above equation gives

$$z = \left(\frac{\omega_0}{s}\right)L = \frac{1}{sC_h} \tag{7.14}$$

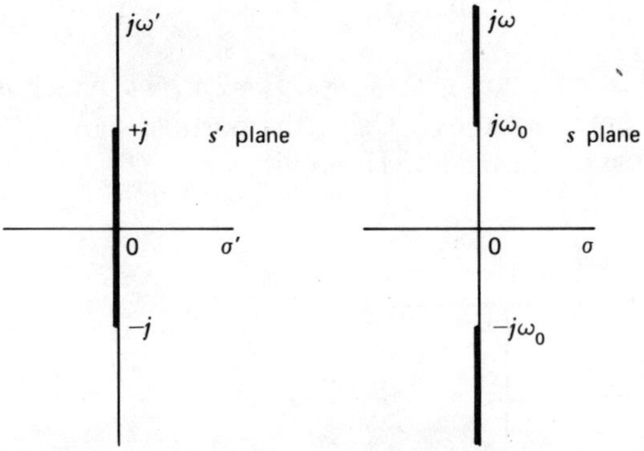

Figure 7.5. Lowpass to highpass transformation.

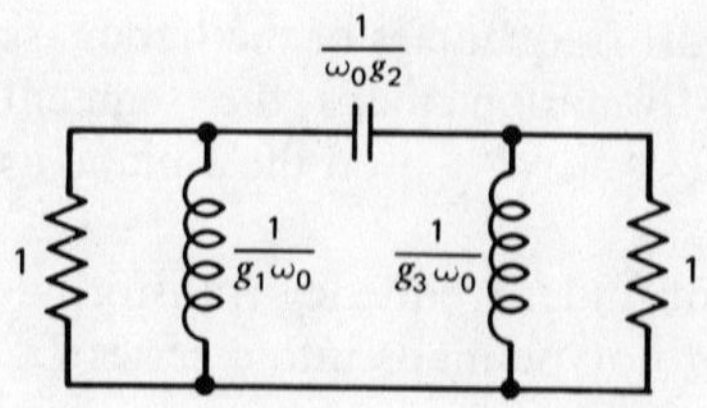

Figure 7.6. ***Highpass prototype, n = 3.***

where

$$C_h = \frac{1}{\omega_0 L} \tag{7.15}$$

Thus the series inductance is transformed into a series capacitance.
For the shunt capacitance the admittance is

$$y = s'C \tag{7.16}$$

Using the frequency transformation gives

$$y = \left(\frac{\omega_0}{s}\right)C = \frac{1}{sL_h} \tag{7.17}$$

where

$$L_h = \frac{1}{\omega_0 C} \tag{7.18}$$

Hence, the shunt capacitance is transformed into a shunt inductance. Figure 7.6 illustrates the $n = 3$ highpass network obtained in this way.

7.5 *LOWPASS TO BANDPASS TRANSFORMATION*

The lowpass to bandpass transformation is given by

$$s' \to \frac{\omega_0}{BW}\left(\frac{s}{\omega_0} + \frac{\omega_0}{s}\right) \tag{7.19}$$

where

$$BW = \omega_{c2} - \omega_{c1} \tag{7.20}$$

$$\omega_0^2 = \omega_{c1}\omega_{c2} \tag{7.21}$$

ω_{c1} and ω_{c2} denote the lower and upper cut-off frequencies of the bandpass filter and BW its bandwidth. This transformation maps the segment $0<|\omega'|<1$ on the s' plane to the segment $|\omega_{c2}|<|\omega|<\omega_{c1}|$ on the s-plane as shown in Figure 7.7.

Using the fact that impedance is invariant under frequency transformation, the series inductance of the lowpass prototype maps into a series LC resonator for the bandpass filter

$$z = s'L = \left[\frac{\omega_0}{BW}\cdot\left(\frac{s}{\omega_0}+\frac{\omega_0}{s}\right)\right]L \tag{7.22}$$

Thus z takes the form

$$z = sL_s + \frac{1}{sC_s} \tag{7.23}$$

where

$$L_s = \frac{L}{BW} \tag{7.24}$$

$$C_s = \frac{BW}{\omega_0^2 L} \tag{7.25}$$

Applying the frequency transformation to the shunt capacitance of the lowpass prototype gives

$$y = s'C = \left[\frac{\omega_0}{BW}\left(\frac{s}{\omega_0}+\frac{\omega_0}{s}\right)\right]C \tag{7.26}$$

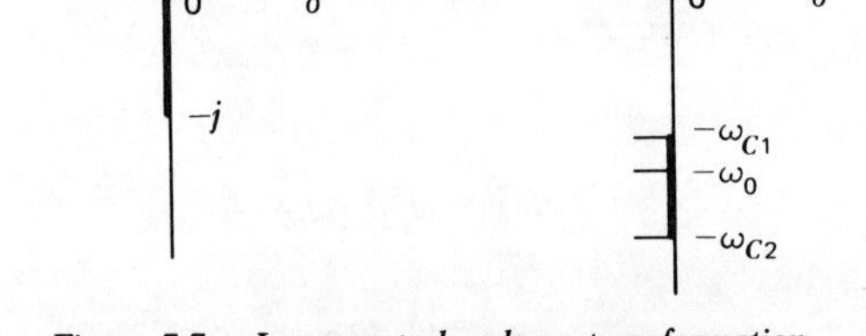

Figure 7.7. Lowpass to bandpass transformation.

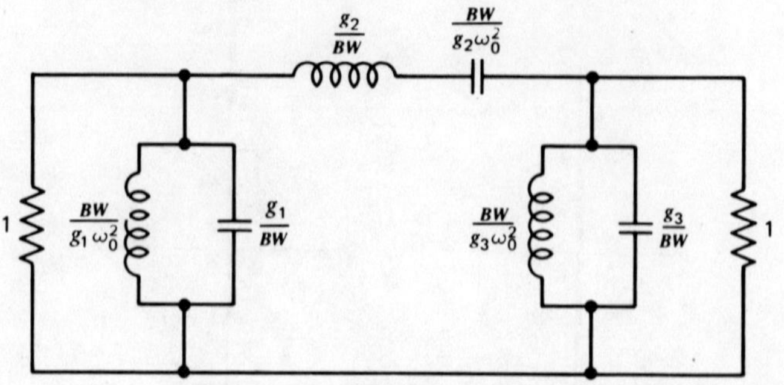

Figure 7.8. Bandpass prototype, n=3.

The form for y is

$$y = sC_p + \frac{1}{sL_p} \tag{7.27}$$

where

$$C_p = \frac{C}{BW} \tag{7.28}$$

$$L_p = \frac{BW}{C\omega_0^2} \tag{7.29}$$

Thus the shunt capacitance maps into a shunt LC circuit. Figure 7.8 depicts the $n=3$ bandpass circuit arrangement.

7.6 *LOWPASS TO BANDSTOP TRANSFORMATION*

The mapping between the lowpass and bandstop filters proceeds in a similar fashion to that used to map the lowpass to the highpass and bandpass ones. This transformation is defined by

$$s' = \frac{BW}{\omega_0\left(\dfrac{s}{\omega_0} + \dfrac{\omega_0}{s}\right)} \tag{7.30}$$

where the variables have the meaning defined earlier. This transformation maps the segment $|\omega'| < 1$ to $|\omega_{c2}| > |\omega| > |\omega_{c1}|$ as indicated in Figure 7.9.

Since immittance is invariant under frequency transformation, the series

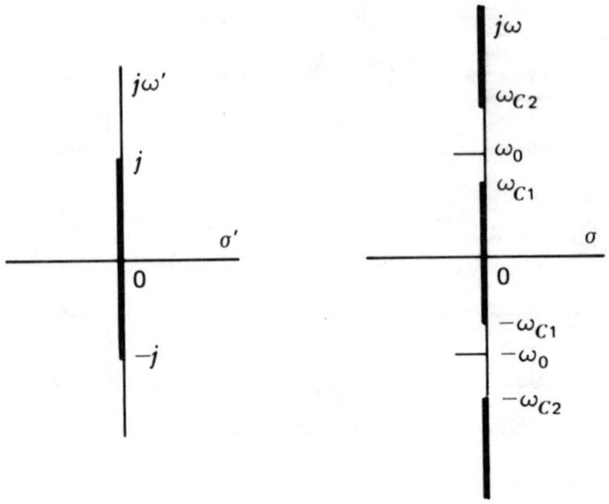

Figure 7.9. Lowpass to bandstop transformation.

inductance maps into a shunt-tuned circuit with element values

$$C_p = \frac{1}{LBW} \tag{7.31}$$

$$L_p = \frac{LBW}{\omega_0^2} \tag{7.32}$$

whereas the shunt capacitance maps into a series-tuned circuit with element values

$$C_s = \frac{CBW}{\omega_0^2} \tag{7.33}$$

$$L_s = \frac{1}{CBW} \tag{7.34}$$

The circuit arrangement for the bandstop filter with $n = 3$ is shown in Figure 7.10.

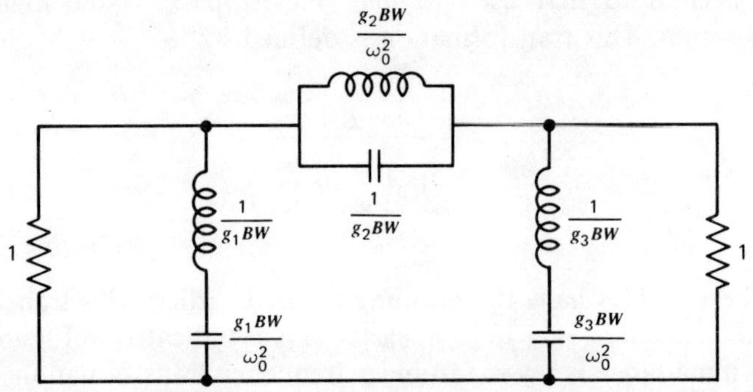

Figure 7.10. Bandstop prototype, $n = 3$.

7.7 FREQUENCY SCALING

Although the various frequency transformations introduced so far have incorporated frequency scaling of the resultant filter networks, the lowpass prototype is still scaled to 1 rad/sec. To obtain an arbitrary cut-off frequency the following frequency transformation is used:

$$s' \to \frac{s}{\omega_0} \tag{7.35}$$

where ω_0 is dimensionless.

Using the fact that immittance remains invariant under frequency transformation gives

$$s'L = \left(\frac{s}{\omega_0}\right)L = sL' \tag{7.36}$$

where

$$L' = \frac{L}{\omega_0} \tag{7.37}$$

Similarly, for the capacitance

$$s'C = \left(\frac{s}{\omega_0}\right)C = sC' \tag{7.38}$$

Thus

$$C' = \frac{C}{\omega_0} \tag{7.39}$$

Resistors obviously remain unaffected by frequency scaling.

7.8 IMPEDANCE SCALING

All the filters that have been constructed have been working into $R = 1$ ohm. The impedance level of the filter can be scaled in the following way. Suppose the actual impedance level should be R_0 ohm instead of 1 ohm. Then a denormalizing impedance Z'' is related to a normalized one Z by

$$Z'' = R_0 Z \tag{7.40}$$

where R_0 is dimensionless. Thus, the normalized resistor R becomes

$$R'' = R_0 R = R_0 \tag{7.41}$$

For an inductance impedance scaling gives L'' as

$$sL'' = R_0(sL) \tag{7.42}$$

Hence

$$L'' = R_0 L \tag{7.43}$$

For a capacitance C'' the result is

$$\frac{1}{sC''} = R_0\left(\frac{1}{sC}\right) \tag{7.44}$$

This gives the scaled capacitance as

$$C'' = \frac{C}{R_0} \tag{7.45}$$

7.9 *IMMITTANCE INVERTERS*

At microwave frequencies it is difficult to lump all branches of the filter at one place. It is therefore necessary to distribute the branch circuits along the transmission line so that mutual immittances become negligible. One practical approach relies on immittance inverters to simulate a ladder network of alternate series and shunt branches by shunt branches only (or series branches only) along the transmission line as shown in Figures 7.11 and 7.12. One immittance inverter is the quarter-wave transmission line of characteristic immittance K_0 or J_0.

The equivalence between the series element in Figure 7.13*a* and the shunt network in Figure 7.13*b* loaded on each side by a quarter-wave immittance inverter may be understood by comparing their transmission coefficients. For the circuit in Figure 7.13*a*, the reflection eigenvalues are

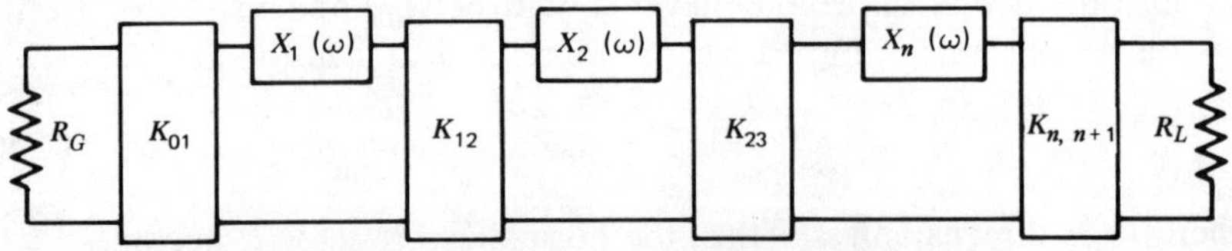

Figure 7.11. Bandpass filter using impedance inverters.

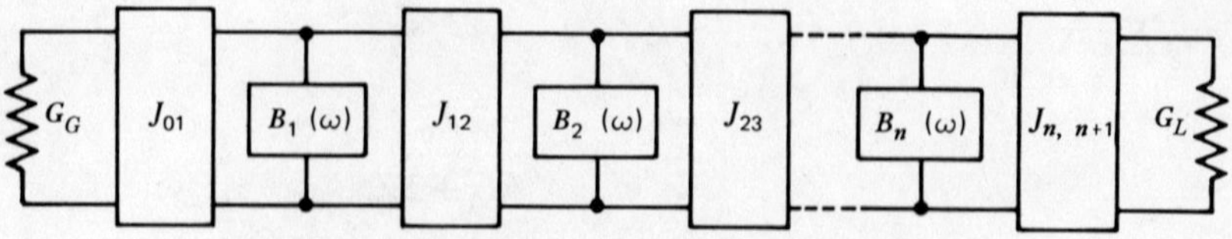

Figure 7.12. *Bandpass filters using admittance inverters.*

given by inspection as

$$s_1 = s_{o/c} = 1 \tag{7.46}$$

$$s_2 = s_{s/c} = \frac{Z/2 - Z_0}{Z/2 + Z_0} \tag{7.47}$$

Thus

$$S_{21} = \frac{s_1 - s_2}{2} = \frac{2Z_0}{Z + 2Z_0} \tag{7.48}$$

For the circuit in Figure 7.13b the reflection eigenvalues are

$$s_1 = s_{o/c} = \frac{Y_0 - 2J_0^2/Y}{Y_0 + 2J_0^2/Y} \tag{7.49}$$

$$s_2 = s_{s/c} = 1 \tag{7.50}$$

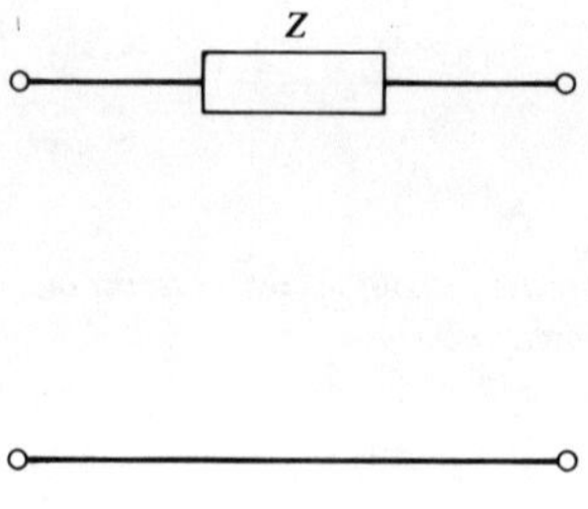

Figure 7.13a. *Series impedance.*

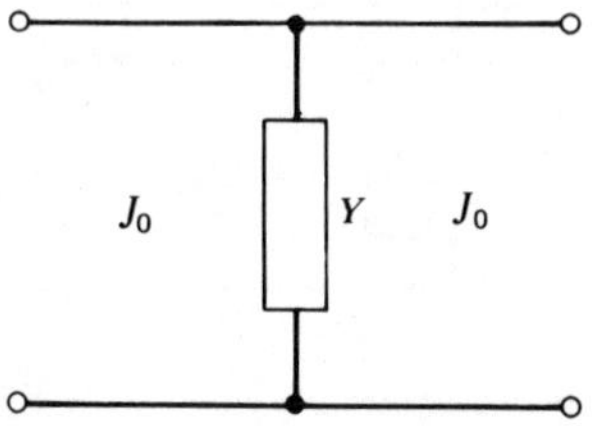

Figure 7.13b. *Series impedance using shunt network and admittance inverters.*

Thus S_{21} is

$$S_{21}=\frac{s_1-s_2}{2}=\frac{-2Z_0}{Y/J_0^2+2Z_0} \tag{7.51}$$

The two networks have therefore similar coefficients provided

$$Y=ZJ_0^2 \tag{7.52}$$

For the dual circuits in Figures 7.14*a* and 7.14*b* the necessary relation is

$$Z=YK_0^2 \tag{7.53}$$

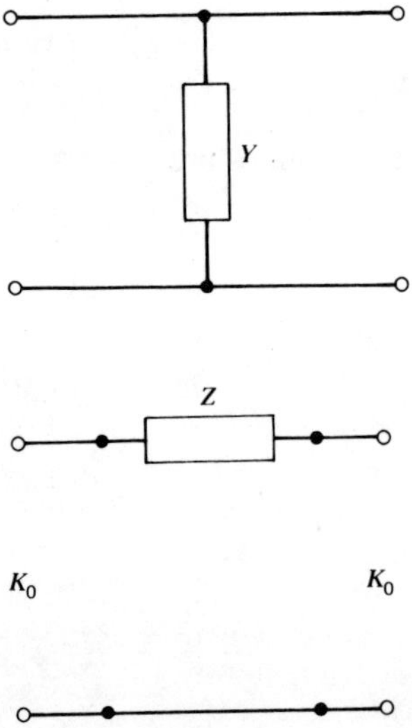

Figure 7.14a. Shunt admittance.

Figure 7.14b. Shunt admittance constructed in terms of series network and impedance inverters.

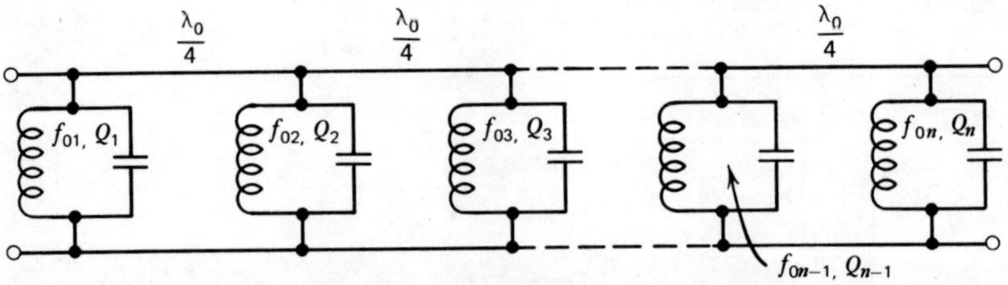

Figure 7.15. Bandpass filter using quarterwave immittance inverters.

Figure 7.15 depicts one bandpass filter circuit that relies on the J immittance inverter described here. The transformations defined by Eq. 7.52 and 7.53 may be applied to the lowpass prototype before further transformations or directly to the final bandpass, bandstop, and highpass filter networks.

7.10 *MICROWAVE BANDPASS FILTER CONFIGURATION*

At microwave frequencies the lumped element resonators in the bandpass and bandstop filters must be replaced by distributed ones. The synthesis of lumped element filters is readily extended to distributed ones by defining the ladder branches in terms of Q factors instead of L's and C's. The two descriptions are equivalent provided their Q factors are identical. Writing the doubly terminated Q factor of the rth branch in terms of the values of L_r and C_r defined by either Eqs. 7.24 and 7.25 or by Eqs. 7.28 and 7.29 gives a single formula for either the shunt or series circuits.

$$Q_r = \frac{g_r \omega_0}{2BW} \tag{7.54}$$

where L_n and C_n have been replaced by the single variable g_r. Using the recurrence formula for g_r given by Eq. 7.11 gives

$$Q_r = \frac{\omega_0 \sin\left[(2r-1)\pi/2n\right]}{BW} \tag{7.55}$$

This last equation is also sometimes written as

$$Q_r = Q_t \sin\frac{(2r-1)\pi}{2n} \tag{7.56}$$

where

$$Q_t = \frac{\omega_0}{BW} \tag{7.57}$$

Once the Q factors of the resonators are known, it is necessary to relate them to the geometry of the microwave circuit. One often used configuration consisting of a half-wave section of transmission line terminated by two susceptances, this is described in Figure 6.14. The final step in the construction of such bandpass waveguide filters is to relate the susceptances of the cavity to the Q factor of the filter. The required relation is given in Chapter 6 by

$$Q_l = \frac{\pi B}{4}\left(\frac{\lambda_g}{\lambda_0}\right)^2 \sqrt{B^2+4}$$

This equation differs from that stated in Eq. 6.69 by the factor $(\lambda_g/\lambda_0)^2$ which applies to a dispersive line, such as a waveguide.

PROBLEMS

1. Using the lowpass to bandpass frequency transformation, show that the transfer function $|S_{21}^2(j\omega)|$ for the bandpass filter takes the form $1/\{1+[Q_t(\omega-\omega_0)/(\omega_0-\omega)]^{2n}\}$.
2. Obtain the element values of the highpass, bandpass, and bandstop filters for $n=3$ using the elements of the lowpass Butterworth prototype.
3. Verify the mappings on the $j\omega$ axis between the s and s_n planes for the lowpass prototype to the highpass, bandpass, and bandstop filter prototypes.
4. Synthesise a lowpass filter for $|S_{21}(j\omega)|^2=1/(1+\omega^4)$.
5. Obtain 7.54 by first impedance scaling g_r
6. Derive the elements of the bandstop filter network from the lowpass prototype network.
7. Construct a lowpass filter with $n=3$ having a cut-off frequency of 1000 rad/sec and 50 Ω terminations.
8. Using the plus sign in Eq. 7.9 synthesise an $n=3$ lowpass T circuit
9. Obtain the element values of the highpass, bandpass, and bandstop filters for the lowpass prototype in problem 8.

REFERENCES

1. S. Darlington, Synthesis of Reactance Four Poles which Produce Prescribed Insertion Loss Characteristics, Including Special Applications to Filter Design, *J. Math. Phys.*, **18**, (1939), p. 257.

FURTHER READING

1. W. W. Mumford, *Maximally Flat Filters in Waveguide*, *BSTJ*, **27**, (1948), p. 684.
2. J. O. Scanlon and R. Levy, *Circuit Theory*, Vol. 2, Oliver and Boyd, Edinburgh, 1973.

CHAPTER EIGHT

Nonreciprocal Ferrite Devices

An important class of microwave components are the nonreciprocal ferrite ones that rely for their operation on the tensor form for the permeability of a magnetized ferrite material. The nature of the permeability tensor is associated with the motion of magnetic dipoles in magnetic insulators in the presence of a constant magnetic field and a superimposed microwave magnetic one. The behavior of microwave nonreciprocal devices rests on the fact that, although the permeability of a magnetized ferrite material is in general a tensor, it is characterized by two different scalar permeabilities for oppositely rotating microwave magnetic fields perpendicular to the direct magnetic field.

The first practical nonreciprocal device was the one-way transmission line using Faraday rotation in a longitudinally magnetized circular waveguide. Two-port microwave devices in a rectangular waveguide include nonreciprocal phase shifters, resonance isolators, ferrite limiters and ferrimagnetic filters. Another important class of ferrite device is the 3-port circulator, in which a wave incident at one port is emergent at a second with none entering the third one.

Although the nonreciprocal components are by far the most important ferrite devices, the dispersive character of the permeability also leads to reciprocal variable phase and attenuator networks.

Much of the macroscopic theory of microwave ferrite devices is based on the equation of motion of the magnetization vector. This equation can be derived by considering an elementary magnetic dipole having a dipole moment $\bar{\mu}$ placed in a d.c. magnetic field $\bar{H}_0$. Under equilibrium conditions, the dipole moment vector $\bar{\mu}$ lies in the direction $\bar{H}_0$, which is usually assumed to be in the z direction. Now let us assume that the magnetic dipole is tilted by a small external force so that it makes an angle θ with $\bar{H}_0$ as shown in Fig. 8.1. Since the only field acting on $\bar{\mu}$ is $\bar{H}_0$ the torque exerted on $\bar{\mu}$ is

$$\bar{T} = \bar{\mu} \times \bar{H}_0 \tag{8.1}$$

Associated with the magnetic dipole $\bar{\mu}$ there is an angular momentum $\bar{J}$ given by

$$\bar{\mu} = \gamma \bar{J} \tag{8.2}$$

where γ is the gyromagnetic ratio given by -2.21×10^5(rad/sec)/(A./m.)

From Eq. 8.2 the torque can also be written as

$$\bar{T} = \frac{1}{\gamma} \frac{d\bar{\mu}}{dt} \tag{8.3}$$

Combining Eqs. 8.1 and 8.3 gives the equation of motion for a single dipole

$$\frac{d\bar{\mu}}{dt} = \gamma \left(\bar{\mu} \times \bar{H}_0 \right) \tag{8.4}$$

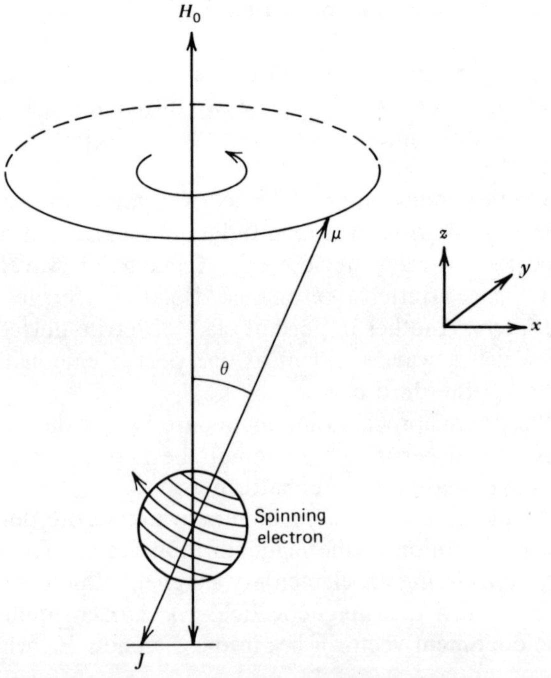

Figure 8.1. Magnetic moment processing about a magnetic field.

The total magnetic moment per unit volume in Webers per square meter is

$$\overline{M}_0 = N\overline{\mu} \tag{8.5}$$

where N is the number of unbalanced spins per unit volume. Equation 8.4 now becomes

$$\frac{d\overline{M}_0}{dt} = \gamma\left(\overline{M}_0 \times \overline{H}_0\right) \tag{8.6}$$

Much of the classical theory of microwave ferrites is based on the equation of motion of the magnetization vector given by Eq. 8.6.

8.1 *SUSCEPTIBILITY TENSOR IN INFINITE MEDIUM*

In the most simple microwave case the total effective magnetic field in Eq. 8.6 consists of the d.c. magnetic field $\overline{H}_0$ and the r.f. magnetic field $\overline{h}$

$$\overline{H} = \overline{H}_0 + \overline{h} \tag{8.7}$$

The total magnetization consists of the d.c. magnetization $\overline{M}_0$ and the r.f. magnetization $\overline{m}$

$$\overline{M} = \overline{M}_0 + \overline{m} \tag{8.8}$$

In component form the above equations are

$$\overline{H}_0 = \begin{bmatrix} 0 \\ 0 \\ H_0 \end{bmatrix} \tag{8.9}$$

$$\overline{h} = \begin{bmatrix} h_x \\ h_y \\ h_z \end{bmatrix} \tag{8.10}$$

$$\overline{M}_0 = \begin{bmatrix} 0 \\ 0 \\ M_0 \end{bmatrix} \tag{8.11}$$

$$\overline{m} = \begin{bmatrix} m_x \\ m_y \\ m_z \end{bmatrix} \tag{8.12}$$

Equation 8.6 can now be expanded as

$$\frac{dm_x}{dt} = m_y\gamma(H_0+h_z) - h_y\gamma(M_0+m_z) \tag{8.13}$$

$$\frac{dm_y}{dt} = -m_x\gamma(H_0+h_z) + h_x\gamma(M_0+m_z) \tag{8.14}$$

$$\frac{dm_z}{dt} = m_x\gamma h_y - m_y\gamma h_x \tag{8.15}$$

In the small signal approximation higher order terms of m and h are set equal to zero. The small signal approximation is therefore

$$\frac{dm_x}{dt} = m_y\gamma H_0 - h_y\gamma M_0 \tag{8.16}$$

$$\frac{dm_y}{dt} = -m_x\gamma H_0 + h_x\gamma M_0 \tag{8.17}$$

$$\frac{dm_z}{dt} \approx 0 \tag{8.18}$$

Rewriting the last equations gives

$$\frac{d^2m_x}{dt^2} + \omega_0^2 m_x = \mu_0\omega_m\omega_0 h_x - \mu_0\omega_m\frac{dh_y}{dt} \tag{8.19}$$

$$\frac{d^2m_y}{dt^2} + \omega_0^2 m_y = \mu_0\omega_m\frac{dh_x}{dt} + \mu_0\omega_m\omega_0 h_y \tag{8.20}$$

$$m_z \approx 0 \tag{8.21}$$

where

$$\omega_m = \frac{\gamma M_0}{\mu_0} \tag{8.22}$$

$$\omega_0 = \gamma H_0 \tag{8.23}$$

If the time dependence of the r.f. quantities is of the form $\exp(j\omega t)$, a susceptibility tensor $[\chi]$ can be defined that relates the r.f. magnetization to the r.f. magnetic field

$$\bar{m} = \mu_0[\chi]\bar{h} \tag{8.24}$$

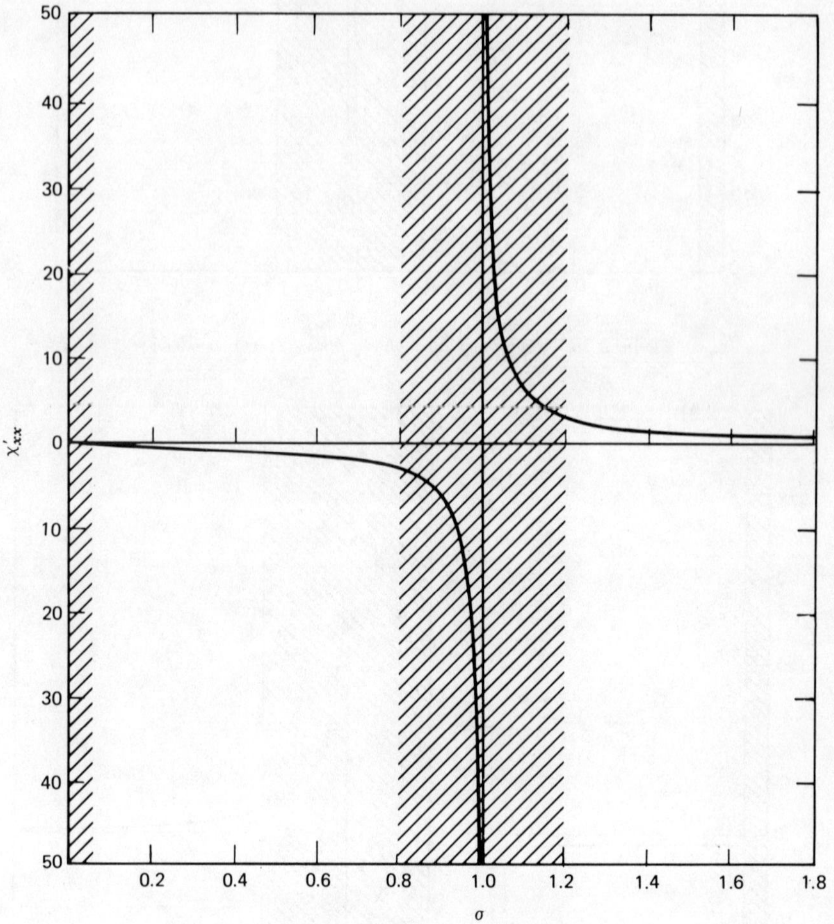

Figure 8.2a. Real part of χ_{xx} for $\omega_m/\omega=1$, $\alpha=0.01$.

where

$$[\chi]=\begin{bmatrix}\chi_{xx} & \chi_{xy} & 0\\ \chi_{yx} & \chi_{yy} & 0\\ 0 & 0 & 0\end{bmatrix} \tag{8.25}$$

and

$$\chi_{xx}=\chi_{yy}=\frac{\omega_m\omega_0}{-\omega^2+\omega_0^2} \tag{8.26}$$

$$\chi_{yx}=-\chi_{xy}=\frac{j\omega_m\omega}{-\omega^2+\omega_0^2}=jK \tag{8.27}$$

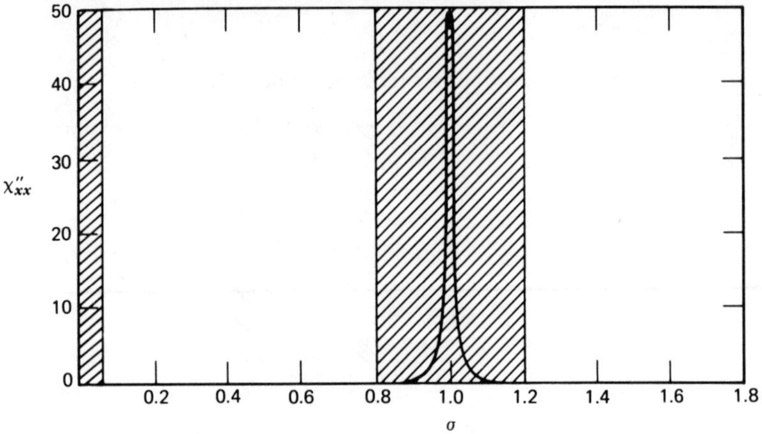

Figure 8.2b. Imaginary part of χ_{xx} for $\omega_m/\omega=1$, $\alpha=0.01$.

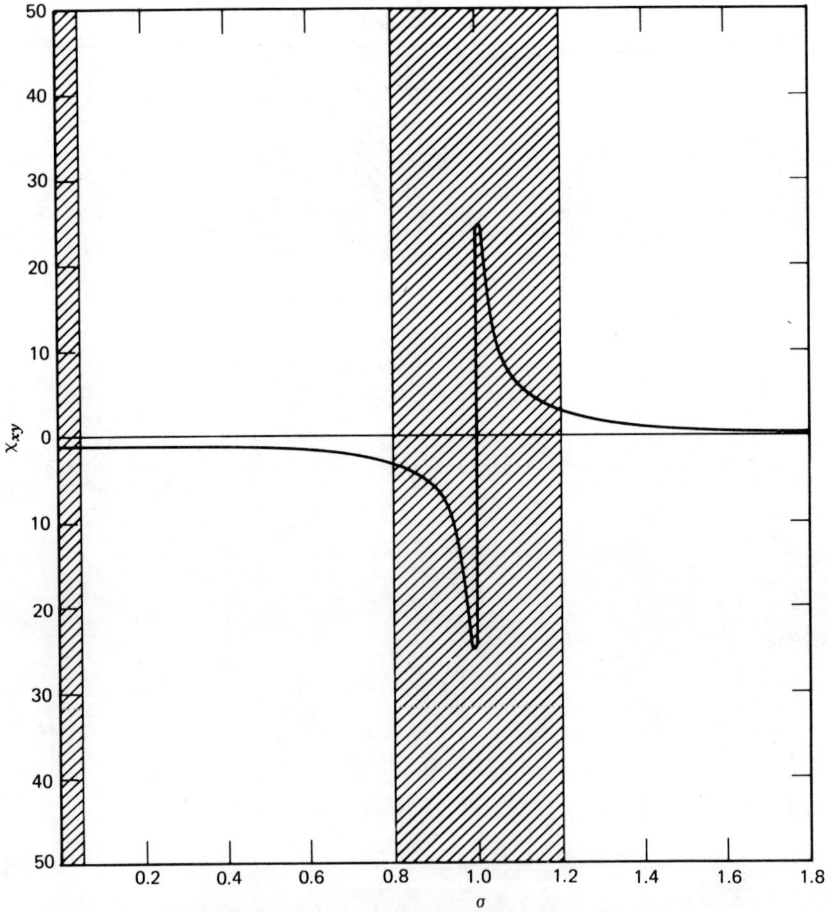

Figure 8.2c. Real part of K for $\omega_m/\omega=1$, $\alpha=0.01$.

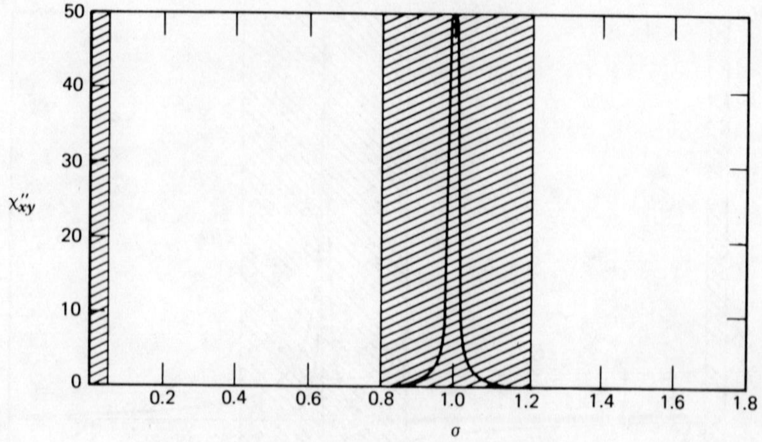

Figure 8.2d. Imagninary part of K for $\omega_m/\omega=1$, $\alpha=0.01$.

The components of the susceptibility tensor have a singularity at $\omega=\omega_0$. This is defined as the resonance condition.

To stabilize the motion of the magnetization vector at the resonance, a damping term must be introduced into the equation of motion. One form of phenomenological damping often used is obtained by replacing ω_0 by $\omega_0+j\alpha$, where α may be related to the effective line width of the ferrite material. Figures 8.2*a* through 8.2*d* depict the real and imaginary parts of χ_{xx} and χ_{xy} for $\omega_m/\omega=1$ and $\alpha=0.01$.

8.2 *SCALAR SUSCEPTIBILITY*

The relation between $\overline{m}$ and $\overline{h}$ is a scalar quantity provided $\overline{h}$ is adjusted to correspond to one of the normal modes of the system. The modes may be found by first finding the eigenvalues (scalar susceptibilities) of the tensor susceptibility. The eigenvalue can be solved as

$$\chi\overline{H}=[\chi]\overline{H} \tag{8.28}$$

where χ is an eigenvalue, $\overline{H}$ is an eigenvector, and $[\chi]$ is the susceptibility tensor. The magnetic fields are proportional to the eigenvectors.

Equation 8.28 has a nonvanishing value for $\overline{H}$ provided

$$\begin{vmatrix} (\chi_{xx}-\chi) & \chi_{xy} & 0 \\ -\chi_{xy} & (\chi_{xx}-\chi) & 0 \\ 0 & 0 & -\chi \end{vmatrix}=0 \tag{8.29}$$

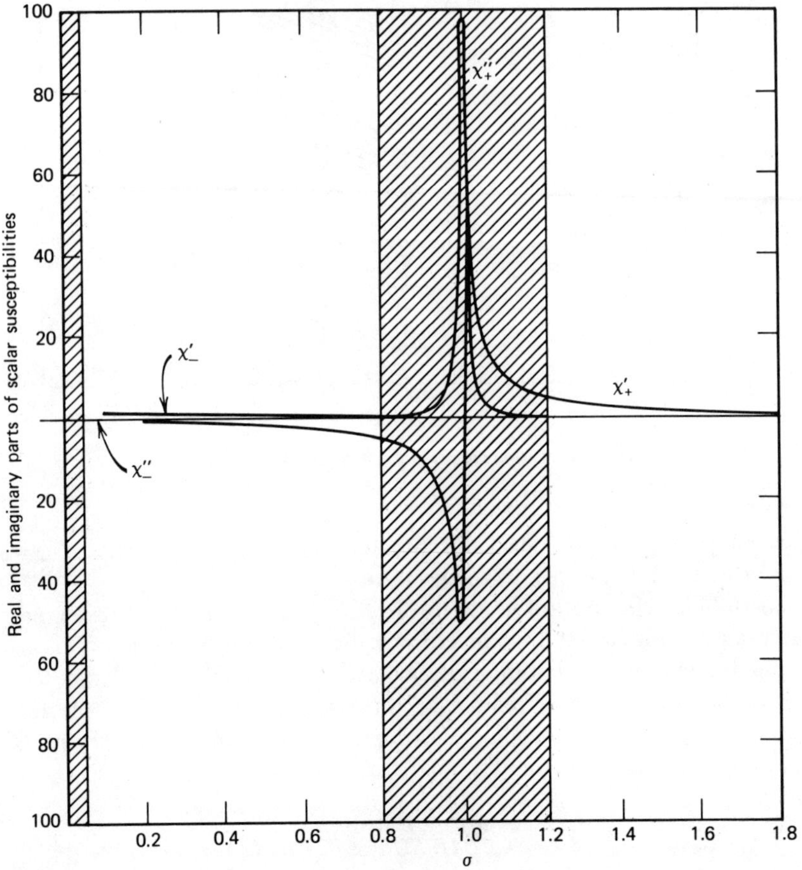

Figure 8.3. Real and imaginary parts of scalar susceptibilities for $\omega_m/\omega=1$, $\alpha=0.01$.

The three eigenvalues are

$$\chi_1=\chi_+=\chi_{xx}+j\chi_{xy} \tag{8.30}$$

$$\chi_2=\chi_-=\chi_{xx}-j\chi_{xy} \tag{8.31}$$

$$\chi_3=0 \tag{8.32}$$

The eigenvectors are now found by substituting Eq. 8.30 through 8.32 into

Eq. 8.28. Each eigenvalue corresponds to one of the eigenvectors.

$$\overline{H}_1=\bar{h}_+=\frac{1}{\sqrt{2}}\begin{bmatrix} h_0 \\ jh_0 \\ 0 \end{bmatrix} \tag{8.33}$$

$$\overline{H}_2=\bar{h}_-=\frac{1}{\sqrt{2}}\begin{bmatrix} h_0 \\ -jh_0 \\ 0 \end{bmatrix} \tag{8.34}$$

$$\overline{H}_3=\bar{h}_z=\begin{bmatrix} 0 \\ 0 \\ h_0 \end{bmatrix} \tag{8.35}$$

The magnetic fields corresponding to the last three equations are orthonormal. Figure 8.3 gives the real and imaginary parts of $\chi_\pm$ for $\omega_m/\omega=1$, $\alpha=0.01$.

8.3 *TENSOR PERMEABILITY*

A tensor permeability can also be defined by relating the r.f. flux density $\bar{b}$ to the r.f. magnetic field $\bar{h}$

$$\bar{b}=\mu_0\bar{h}+\bar{m} \tag{8.36}$$

or

$$\bar{b}=\mu_0[\,\mu\,]\bar{h} \tag{8.37}$$

where

$$[\,\mu\,]=[1]+[\chi] \tag{8.38}$$

Using Eq. 8.25 the tensor permeability is

$$[\,\mu\,]=\begin{bmatrix} \mu & -jK & 0 \\ jK & \mu & 0 \\ 0 & 0 & 1 \end{bmatrix} \tag{8.39}$$

where

$$\mu=1+\chi_{xx} \tag{8.40}$$

$$jK=-\chi_{xy} \tag{8.41}$$

8.4 *SCALAR PERMEABILITY*

The permeability is also a scalar for the normal r.f. magnetic fields defined by Eqs. 8.33 through 8.35

$$\mu_1 = \mu_+ = \mu - K \tag{8.42}$$

$$\mu_2 = \mu_- = \mu + K \tag{8.43}$$

$$\mu_3 = \mu_z = 1 \tag{8.44}$$

This result is obtained quite simply by forming the eigenvalues of the permeability tensor given by Eq. 8.39. The damping can again be introduced into Eqs. 8.42 and 8.43 by replacing ω_0 by $\omega_0 + j\alpha$. In keeping with the behavior of the scalar susceptibilities, the positive scalar permeability has a singularity at $\omega = \omega_0$, whereas the negative scalar susceptibility does not exhibit a resonance there.

8.5 *FARADAY ROTATION*

The most simple problem involving the tensor permeability is that of propagation in an infinite medium. This simple introduction serves to demonstrate the well-known Faraday rotation effect on which a number of different ferrite devices depend. The arrangement considered is that where the direction of propagation and the direct magnetizing magnetic field are in the same direction.

The wave equation is from Maxwell's equations given by

$$\nabla \times \nabla \times \bar{h} = \omega^2 \epsilon_0 \epsilon_r \mu_0 [\mu] \bar{h} \tag{8.45}$$

where

$$[\mu] = \begin{bmatrix} \mu & -jK & 0 \\ jK & \mu & 0 \\ 0 & 0 & 1 \end{bmatrix} \tag{8.46}$$

$$\bar{h} = \begin{bmatrix} h_x \\ h_y \\ 0 \end{bmatrix} \tag{8.47}$$

Assuming propagation of the form $\exp(-\gamma z)$ with no variation of the fields in the transverse x-y plane.

$$\nabla \times \nabla \times \Rightarrow \frac{\partial^2}{\partial z^2} \Rightarrow \gamma^2 \tag{8.48}$$

Combining the above equations gives

$$\gamma^2\begin{bmatrix} h_x \\ h_y \end{bmatrix} = \omega^2\epsilon_0\epsilon_r\mu_0\begin{bmatrix} \mu & -jK \\ jK & \mu \end{bmatrix}\begin{bmatrix} h_x \\ h_y \end{bmatrix} \tag{8.49}$$

The roots of this equation are

$$\begin{vmatrix} \mu - \dfrac{\gamma^2}{\omega^2\epsilon_r\epsilon_0\mu_0} & -jK \\ jK & \mu - \dfrac{\gamma^2}{\omega^2\epsilon_r\epsilon_0\mu_0} \end{vmatrix} = 0 \tag{8.50}$$

Thus

$$\gamma_{\pm}^2 = \omega^2\epsilon_0\epsilon_r\mu_0(\mu \mp K) \tag{8.51}$$

The two field patterns corresponding to $\gamma_{\pm}^2$ are now obtained by introducing $\gamma_{\pm}$ one at a time into Eq. 8.49. The result is

$$h_y^{\pm} = \pm jh_x^{\pm} \tag{8.52}$$

This means that one solution consists of two plane circularly polarized waves rotating in opposite directions with propagation constants $\gamma_{\pm}$.

To obtain the behavior of a linearly polarized wave in a magnetized ferrite medium, it is sufficient to take a linear combination of the two circularly polarized waves defined by Eq. 8.52.

$$h_x(z) = \frac{1}{2}\exp(-j\beta_- z) + \frac{1}{2}\exp(-j\beta_+ z) \tag{8.53}$$

$$h_y(z) = \frac{j}{2}\exp(-j\beta_- z) - \frac{j}{2}\exp(-j\beta_+ z) \tag{8.54}$$

where for a loss-free medium $\gamma_{\pm}$ is replaced by $j\beta_{\pm}$.

The preceding equations satisfy the boundary conditions $h_x(0)=1$, $h_y(0)=0$. Equations 8.53 and 8.54 may be simplified by taking out a common factor

$$\exp -j\left(\frac{\beta_+ + \beta_-}{2}\right)z$$

This gives

$$h_x(z)=\cos\left(\frac{\beta_+-\beta_-}{2}\right)z\cdot\exp-j\left(\frac{\beta_++\beta_-}{2}\right)z \tag{8.55}$$

$$h_y(z)=\sin\left(\frac{\beta_+-\beta_-}{2}\right)z\cdot\exp-j\left(\frac{\beta_++\beta_-}{2}\right)z \tag{8.56}$$

Combining the above equations gives a linearly polarized wave rotating with the quantity $[(\beta_+-\beta_-)/2]z$. The direction of polarization θ is given by

$$\theta=\left(\frac{\beta_+-\beta_-}{2}\right)z \tag{8.57}$$

If the direction of the direct magnetic field or the direction of propagation is reversed, the angle θ is reversed also.

This result indicates that a wave propagating a certain distance in one direction is rotated through an angle θ with respect to the direct magnetic field. When the wave is reflected and returns to its starting point the rotation is again θ. The total rotation of the reflected wave is therefore 2θ with respect to the outgoing wave, i.e., it is not rotated back to its original orientation. Thus Faraday rotation is nonreciprocal and leads to a number of nonreciprocal devices.

8.6 *FARADAY ROTATION ISOLATOR*

The operation of this type of isolator can be understood with the help of Figure 8.4. The rotator prototype is matched to rectangular waveguide at the two ends by quarter-wave transformers in round waveguide. Resistance cards are inserted in the round waveguide sections in a plane perpendicular to the electric fields of the input and output rectangular waveguides. A signal incident at the input will be perpendicular to the first resistance card, and after a clockwise rotation through 45 deg, will also be perpendicularly polarized with respect to the output resistance card. The signal, therefore, will be transmitted without attenuation through the isolator. In the reverse direction, a signal incident on the output resistance card is likewise perpendicular to that card, but after rotation through 45 deg in a clockwise sense, will now be in the plane of the input resistance card where it is attenuated.

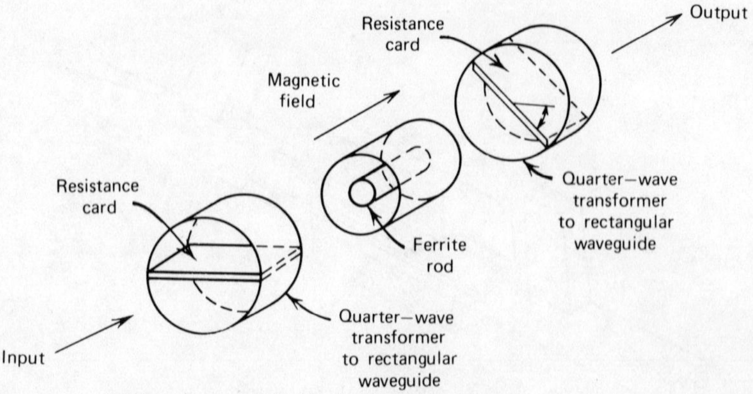

Figure 8.4. Schematic diagram of Faraday rotator isolator.

8.7 *FOUR-PORT FARADAY ROTATOR CIRCULATOR*

Another important device based on the Faraday rotator section is the 4-port Faraday rotation circulator. A schematic of a 4-port circulator is shown in Figure 8.5. In this device power entering port 1 emerges from port 2, and so on in a cyclic manner. The physical arrangement is similar to the Faraday rotator isolator except that the sections containing the resistance cards are replaced by two mode transducers as shown in Figure 8.6. The two mode transducers allow orthogonal linearly polarized waves to be applied to the round waveguide section. The Faraday rotator is again a 45-deg section. A wave entering port 1 with its electric field vertically polarized is rotated clockwise by 45 deg by the ferrite rotator and emerges at port 2. A wave entering port 2 is also rotated 45 deg clockwise, so that

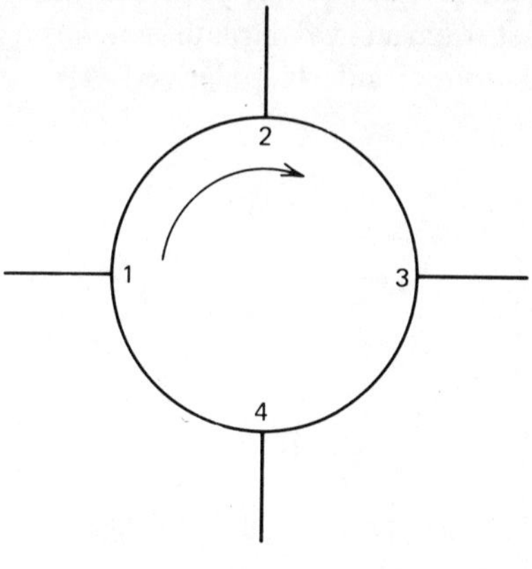

Figure 8.5. Schematic diagram of 4-port circulator.

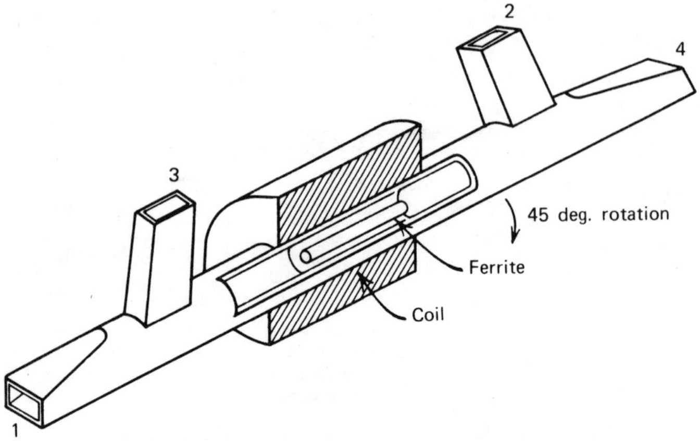

Figure 8.6. Four-port Faraday rotation circulator.

its electric field is now horizontally polarized at the input of the first two mode transducer, and the wave therefore emerges from port 3. Similarly, transmission occurs from port 3 to port 4, port 4 to port 1, and so on.

8.8 *NONRECIPROCAL FARADAY ROTATION TYPE PHASE SHIFTER*

Equations 8.51 and 8.52 suggest that, if the input excitation to the Faraday rotation section corresponds to one of the two normal modes of the system, the wave travels in the same normal mode through the rotator section. This principle can be used to design nonreciprocal or reciprocal ferrite phase shifters in conjunction with reciprocal or nonreciprocal quarter-wave plates. The nonreciprocal phase shifter is shown in Fig. 8.7. The arrangement uses two reciprocal quarter-wave plates at either end of the Faraday rotation section. The first quarter-wave plate converts a linearly polarized input wave into a positive circularly polarized wave at

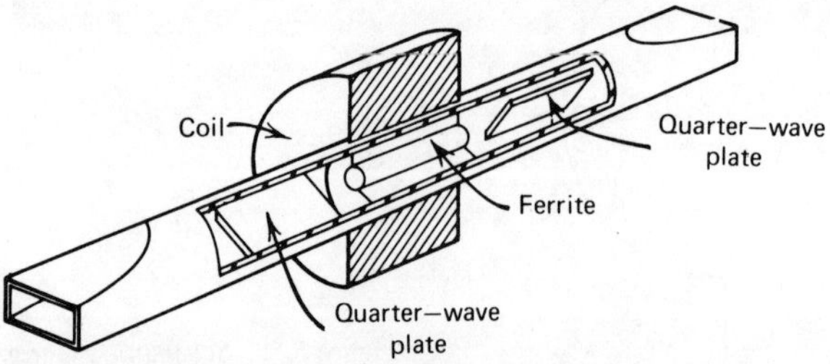

Figure 8.7. Nonreciprocal Faraday rotation phase shifter.

the input of the rotator section. This wave is then phase shifted through $\beta_+ z$ radians by the rotator section. The phase-shifted circularly polarized wave is reconverted to a linearly polarized wave at the output by the second quarter-wave plate. In the reverse direction of propagation, the circularly polarized wave is in the opposite sense and the wave is therefore phase shifted through $\beta_- z$. Hence the arrangement behaves as a nonreciprocal phase shifter. A forward wave can of course also be switched from $\beta_+ z$ to $\beta_- z$ by reversing the d.c. magnetic field on the rotator section.

8.9 *CIRCULAR POLARIZATION IN RECTANGULAR WAVEGUIDE*

Natural regions of circular polarization are also to be found in rectangular waveguide propagating the dominant TE_{10} mode. This feature is particularly important in that a number of ferrite nonreciprocal devices rely on their operation on such polarization. This can be demonstrated very simply by first putting down the three field components for the waveguide.

$$H_z = \cos\frac{\pi x}{a}\cdot\exp\left(-\frac{j2\pi z}{\lambda_g}\right) \tag{8.58}$$

$$H_x = j\left(\frac{\lambda_c}{\lambda_g}\right)\sin\frac{\pi x}{a}\cdot\exp\left(-\frac{j2\pi z}{\lambda_g}\right) \tag{8.59}$$

$$E_y = -j\left(\frac{\lambda_c}{\lambda_0}\right)\sqrt{\frac{\mu_0}{\epsilon_0}}\cdot\sin\frac{\pi x}{a}\cdot\exp\left(-\frac{j2\pi z}{\lambda_g}\right) \tag{8.60}$$

where propagation is assumed along the positive z direction. If propagation is in the negative z direction the sign of H_z is reversed.

In the above equations

$$\lambda_c = 2a \tag{8.61}$$

$$= \sqrt{\frac{\mu_0}{\epsilon_0}} = 120\pi \tag{8.62}$$

$$\beta^2 = \left(\frac{2\pi}{\lambda_g}\right)^2 = \left(\frac{2\pi}{\lambda_0}\right)^2 - \left(\frac{2\pi}{\lambda_c}\right)^2 \tag{8.63}$$

It is observed from the above equations that H_x and H_z are in time-space quadrature. If a region can now be found where their amplitudes are also equal, this region would exhibit circular polarization. This condition is satisfied on either side of the center line of the waveguide provided

$$\tan\left(\frac{\pi x}{a}\right)=\frac{\lambda_g}{\lambda_c} \tag{8.64}$$

The polarization along such a waveguide is in opposite senses on either side of the center line of the waveguide. Furthermore, the two polarizations are reversed provided propagation is taken along the negative z axis instead of the positive one.

8.10 *NONRECIPROCAL PROPAGATION IN RECTANGULAR WAVEGUIDE*

Since the rectangular waveguide operating in its dominant mode supports planes of circular polarization on either side of its center line, it will exhibit nonreciprocal propagation if ferrite slabs are located there. The two propagation constants in such a ferrite-loaded rectangular waveguide may be readily obtained by assuming that the ferrite slabs are sufficiently small so that perturbation theory may be employed. The perturbation formula that applies to the geometry in Figure 8.8 has been given by Lax and Button[1]

$$\Gamma+\Gamma_0^*=\frac{j\omega\int_{\Delta s}\left[\epsilon_0\left(\left[\chi_e^e\right]\bar{E}_0\right)\bar{E}_0^*+\mu_0\left(\left[\chi^e\right]\cdot\bar{h}_0\right)\bar{h}_0^*\right]ds}{2\int_s \bar{i}_y\cdot\left(\bar{E}_0\times\bar{h}_0^*\right)ds} \tag{8.65}$$

In using this formula propagation is taken along the y axis since the applied direct magnetic field is conventionally directed along the z axis, which in this instance is transverse to the direction of propagation.

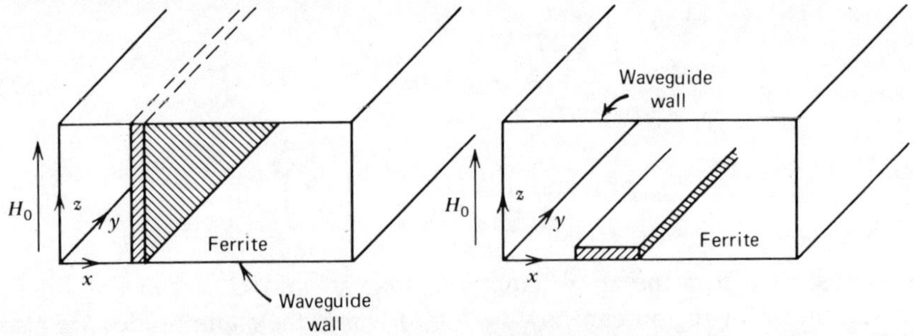

Figure 8.8. Ferrite geometries in rectangular waveguide.

In Eq. 8.65 $\bar{E}_0$ and $\bar{h}_0$ refer to the waveguide fields before the introduction of the ferrite material

$$\bar{E}_0 = \begin{bmatrix} 0 \\ 0 \\ \dfrac{-j\omega\mu_0}{k}\sin kx \end{bmatrix} e^{\mp\Gamma_0 y} \tag{8.66}$$

$$\bar{h}_0 = \begin{bmatrix} \dfrac{\Gamma_0}{k}\sin kx \\ \pm\cos kx \\ 0 \end{bmatrix} e^{\mp\Gamma_0 y} \tag{8.67}$$

and

$$\begin{aligned} -\Gamma_0^2 &= \left(\frac{2\pi}{\lambda_0}\right)^2 - \left(\frac{2\pi}{\lambda_c}\right)^2 \\ &= \left(\frac{2\pi}{\lambda_0}\right)^2 - k^2 \end{aligned} \tag{8.68}$$

The external susceptibility tensor in Eq. 8.65 is given by

$$[\chi^e] = \begin{bmatrix} \chi_{xx}^e & \chi_{xy}^e & 0 \\ -\chi_{xy}^e & \chi_{yy}^e & 0 \\ 0 & 0 & 0 \end{bmatrix} \tag{8.69}$$

The external susceptibility tensor $[\chi^e]$ differs from the internal one in Eq. 8.25 in that its entries relate the internal magnetization to the external magnetic field components in the empty waveguide instead of the fields inside the ferrite as in Eq. 8.25. $[\chi^e]$ is readily obtained by replacing h_x, h_y, and H_0 in Eqs. 8.9 and 8.10 by

$$h_x = h_{0x} - \frac{N_x m_x}{\mu_0} \tag{8.70}$$

$$h_y = h_{0y} - \frac{N_y m_y}{\mu_0} \tag{8.71}$$

$$H_0 = H_{0z} - \frac{N_z M_0}{\mu_0} \tag{8.72}$$

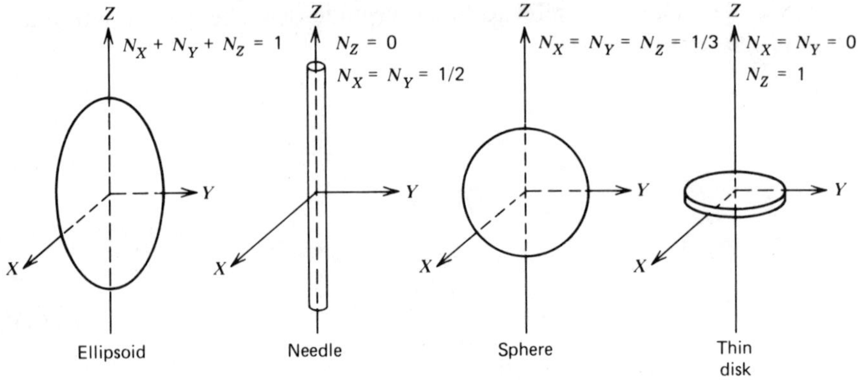

Figure 8.9. Demagnetising factors for some simple ferrite shapes.

The transverse demagnetizing factors N_x and N_y and the z directed one N_z satisfy

$$N_x + N_y + N_z = 1 \tag{8.73}$$

The demagnetizing factors for some simple shapes are illustrated in Figure 8.9. If $N_x \approx N_y$

$$\chi_{xx}^e = \chi_{yy}^e \tag{8.74}$$

in Eq. 8.69.

S and ΔS are the cross-sectional areas of the waveguide and ferrite shapes, χ_e^e is the dielectric constant of the ferrite material. Integrating each integral separately in Eq. 8.65 gives

$$\Gamma_{\pm} - j\beta_0 = j\frac{\Delta S}{S\beta_0}\left[\omega^2\epsilon_0\mu_0\chi_e^e \sin^2 kx + \beta_0^2\chi_{xx}^e \sin^2 kx\right.$$
$$\left. + k^2\chi_{yy}^e \cos^2 kx \pm j\chi_{xy}^e k\beta_0 \sin 2kx\right] \tag{8.75}$$

The $\pm$ signs indicate that the propagation is in general nonreciprocal.

8.11 *PERTURBATION THEORY OF NONRECIPROCAL RECTANGULAR WAVEGUIDE PHASE SHIFTER*

The nonreciprocal phase shift in Eq. 8.75 can be obtained by putting $\Gamma_0 = \alpha + j\beta_0$, and $\chi = \chi' - j\chi''$, and taking its imaginary parts.

$$\beta_{\pm} = \beta_0 + \omega^2\epsilon_0\mu_0\frac{\Delta S}{S\beta_0}(\chi_e^e)' \sin^2 kx$$
$$+ \frac{\Delta S}{S\beta_0}\left[(\chi_{xx}^e)'\beta_0^2 \sin^2 kx + (\chi_{yy}^e)' k^2 \cos^2 kx \pm (\chi_{xy}^e)' k\beta_0 \sin 2kx\right] \tag{8.76}$$

The differential phase shift is therefore given by

$$\beta_+ - \beta_- = -2\frac{\Delta S}{S}\left[(\chi_{xy}^e)' k \sin 2kx\right] \tag{8.77}$$

This equation states that the maximum nonreciprocal phase shift occurs at $x = a/4$.

8.12 *PERTURBATION THEORY OF THE RECTANGULAR WAVEGUIDE RESONANCE ISOLATOR*

One of the most useful of the nonreciprocal ferrite devices is the resonance isolator constructed in the rectangular waveguide. The principle of this device relies on the nonreciprocal attenuation in a ferrite-loaded rectangular waveguide. This can be readily demonstrated by taking the real parts of Eq. 8.75. The result is

$$\alpha_{\pm} = \frac{\Delta S}{S}\left[\beta_0^2(\chi_{xx}^e)'' \sin^2 kx + k^2(\chi_{yy}^e)'' \cos^2 kx \pm 2(\chi_{xy}^e)'' k\beta_0 \sin kx \cos kx\right] \tag{8.78}$$

where we have neglected the dielectric loss χ_e''. The maximum ratio R is

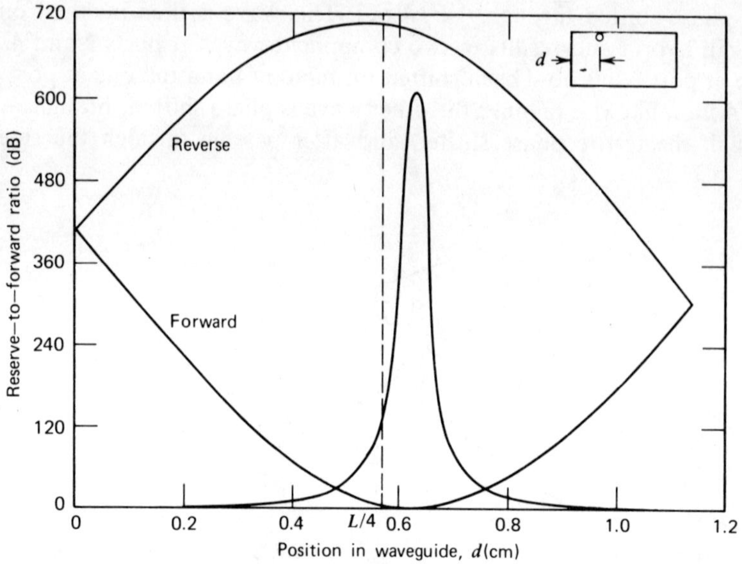

Figure 8.10. Theoretical attenuation using perturbation theory for resonance isolator (after Lax).

given by

$$R = \frac{\chi''_+}{\chi''_-} \tag{8.79}$$

The position at which Eq. 8.79 applies is given from Eq. 8.64 by

$$\tan kx = \frac{k}{\beta_0} \tag{8.80}$$

This last equation gives the position where the r.f. magnetic field is circularly polarized. Figure 8.10 depicts a typical result.

8.13 *DIFFERENTIAL PHASE SHIFT CIRCULATOR*

One application of the differential ferrite phase shifter is in the construction of the differential phase-shift circulator. The differential phase-shift circulator consists of a folded magic T and a 3-dB sidewall hybrid between which is placed a dual section of waveguide containing nonreciprocal 45 deg ferrite phase shifters. The phase shifters in each waveguide are oppositely magnetized. Hence, the total differential phase shift is 90 deg. The operation of this type of circulator can be understood with the help of Figure 8.11. A wave incident on port 1, at the H arm of the magic T, divides equally in phase between the two waveguides. One of the waves is then phase shifted through $+\pi/4$ rad. This wave is then incident on the sidewall hybrid and results in two component waves at ports 2 and 4. The wave at port 2 has now been shifted through $\pi/4$ and the one at port 4 by $3\pi/4$. In a likewise manner the other wave is phase shifted through $-\pi/4$ through the ferrite phase shifter, and after passing through the sidewall

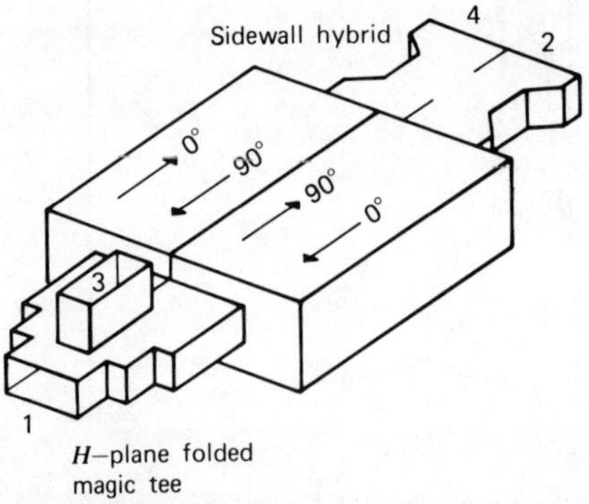

Figure 8.11. Four-port differential phase shift waveguide circulator.

Figure 8.12. x-band waveguide differential phase shift circulator (courtesy of Ferranti Ltd.).

hybrid, gives two more component waves at ports 2 and 4. The one at port 2 is phase shifted through $+\pi/4$ and the one at port 4 through $-\pi/4$. When we combine the two waves at each port, we note that the waves at port 4 are π radians out of phase and, therefore, cancel. On the other hand, the two waves at port 2 are in phase, and, therefore, add. Hence, transmission is from port 1 to port 2. In a similar way transmission occurs from port 2 to port 3, and so on, in a cyclic manner. Figure 8.12 depicts an X band commercial circulator.

PROBLEMS

1. Construct the $\bar{S}$ matrix of a 45-deg Faraday rotation section.
2. Show that $\mu_{\pm} = 1 + \chi_{\pm}$.
3. Investigate the principle of nonreciprocal phase shift using a Faraday rotation section between two quarter-wave plates.
4. Construct the real and imaginary parts of χ_{xx} and χ_{xy} by replacing ω_0 by $\omega_0 + j\alpha$.
5. Plot $\mu_{\pm}$ for $\omega_m/\omega = 1$, $\alpha = 0.01$.
6. Plot $\mu_{\text{eff}} = (\mu^2 - k^2)/\mu$ for $\omega_m/\omega = 1$, $\alpha = 0.01$.

7. Obtain the entries of the external susceptibility tensor in Eq. 8.69 by replacing the internal fields in Eqs. 8.9 and 8.10 by the external ones in Eqs. 8.70 through 8.72. Omit h_z in this derivation.
8. Show that the external susceptibilities in problem 7 exhibit a resonance at

$$\gamma\{[H_0-(N_z-N_x)M_0/\mu_0][H_0-(N_z-N_y)M_0/\mu_0]\}^{1/2}$$

instead of γH_0 for the internal susceptibilities.
9. The demagnetizing factors of a ferrite sphere are $N_x = N_y = N_z = 1/3$. Find the resonant frequency using the result in the last example. Discuss this result.
10. Show that the propagation constants for a plane wave propagating in an infinite medium magnetized perpendicular to the direction of propagation is

$$\beta=\omega\sqrt{\mu_0\epsilon} \quad \text{and} \quad \beta=\omega\sqrt{(\mu^2-K^2)\epsilon/\mu}$$

11. Obtain the differential phase shift per unit length at 10 GHz of a ferrite 1 mm×2 mm in cross section located 5.715 mm from one side of a waveguide wall with internal dimensions 10.16 mm×22.86 mm. Assume a value of 0.50 for $(\chi_{xy})'$. Calculate the length of a 90 deg differential phase-shift section.
12. Show that

$$\alpha_{\pm}=\frac{2\Delta Sk^2\cos^2 kx}{S}(\chi_{\pm})''$$

at $\tan kx = k/\beta$ provided $\chi_{xx}=\chi_{yy}$. Obtain $\alpha_{\pm}$ for this condition for the geometry in problem 11 for $(\chi_+)''=100$, $(\chi_-)''=0.50$. Calculate the length of a resonance isolator with 10-dB isolation.

REFERENCES

1. B Lax and K. J. Button, *Microwave Ferrites and Magnetics*, McGraw-Hill, New York, 1962.

FURTHER READING

1. R. F. Soohoo, *Theory and Application of Ferrites*, Prentice-Hall, New Jersey, 1960.
2. J. Helszajn, *Principles of Microwave Ferrite Engineering*, Wiley, New York, 1969.
3. J. Helszajn, *Nonreciprocal Junctions and Circulators*, Wiley, New York, 1975.

4. P. J. Clarricoats, *Microwave Ferrites*, Chapman and Hall, London, 1961.

5. R. A. Waldron, *Ferrites—An Introduction for Microwave Engineers*, The Marconi Series, Van Nostrand, New York, 1967.

6. A. G. Fox, S. E. Miller, and M. T. Weiss, Behaviour and applications of ferrites in the microwave region, *Bell Syst. Tech. J.*, **34** (1955), pp. 5–103

7. C. L. Hogan, The ferromagnetic Faraday effect at microwave frequencies and its applications—the microwave gyrator, *Bell Syst. Tech. J.*, **31**, 1952, pp. 22–26.

CHAPTER NINE

YIG Filters

Yttrium iron garnet (YIG) fabricated into highly polished spheres or disks is a high-Q microwave resonator. Such a resonator is particularly useful because it can be tuned over large frequency bands by a direct magnetic field. A microwave resonance occurs in such a resonator when the microwave frequency coincides with the electron precession frequency. The most practical geometry for a YIG resonator is a sphere, because it is easily orientated in the magnetic field and the resonant field is not strongly dependent upon its orientation. The diameter of the sphere is usually between 0.40 and 1.0 mm. The Q factor of this type of resonator is comparable or superior to that of most microwave devices and its small size contrasts dramatically with the equivalent mechanically tuned cavity. Figure 9.1 depicts a raw YIG crystal and some highly polished spheres.

Figure 9.2 illustrates an idealized bandpass filter. Two coils have their axes at right angles to each other, and a small ferrite sphere is placed at the intersection of the coil axes. When the resonator is not magnetized, no power is transferred between the coils because the loop axes are perpendicular to each other and there is no interaction with the ferrite. When a d.c. field is applied along the z axis, the two coils are coupled through the transverse components of the dipolar field of the ferrite resonator. This coupling is largest at ferrimagnetic resonance. The orthogonal circuits can be crossed wires, loops, striplines, waveguides, or cavity modes. Figure 9.3 shows a photograph of the internal microwave structure of a three-stage YIG bandpass filter.

Since the YIG filter is closely related to the nonreciprocal 2-port gyrator network the chapter starts with the definition of this network in terms of its S, Z, and Y matrices and their eigenvalues.

Figure 9.1. Raw YIG crystal and some highly polished spheres (courtesy Watkin-Johnson).

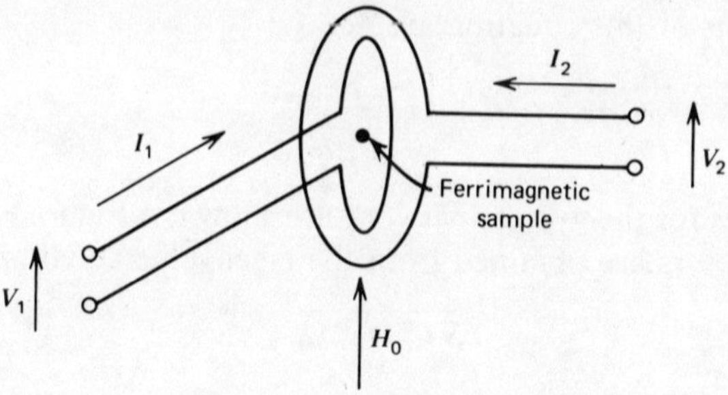

Figure 9.2. Idealized loop coupled YIG resonator.

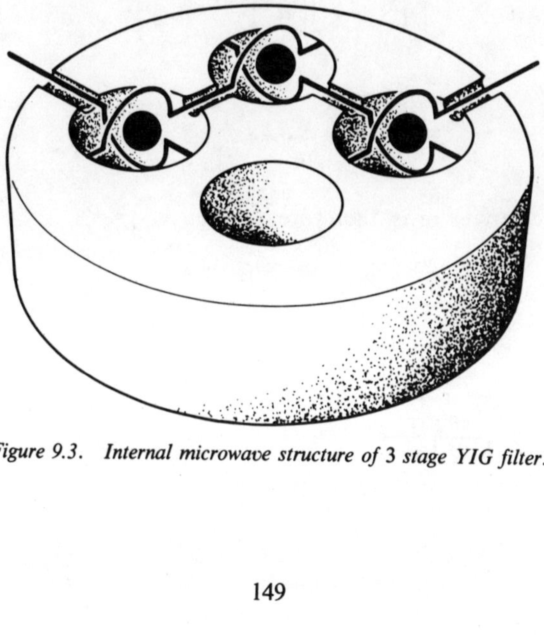

Figure 9.3. Internal microwave structure of 3 stage YIG filter.

9.1 *SCATTERING MATRIX OF 2-PORT GYRATOR NETWORK*

The 2-port gyrator network has a scattering matrix given by

$$\bar{S}=\begin{bmatrix} 0 & -1 \\ 1 & 0 \end{bmatrix} \tag{9.1}$$

This matrix describes the basic nonreciprocal 2-port network defined by Tellegen.[1] Such a network exhibits 180-deg nonreciprocal shift between its input and output terminals. The schematic for this network is shown in Figure 9.4.

The characteristic equation for the above matrix is

$$s_n^2+1=0 \tag{9.2}$$

The two roots of this equation are

$$s_+=j \tag{9.3}$$

$$s_-=-j \tag{9.4}$$

The eigenvalues for the gyrator network are shown in Figure 9.5.

The eigenvectors are obtained from the eigenvalue equation

$$\bar{S}\,\bar{U}_n=s_n\bar{U}_n \tag{9.5}$$

This equation is illustrated in Figure 9.6*a*.

For s_{+1} one has

$$\begin{bmatrix} 0 & -1 \\ 1 & 0 \end{bmatrix}\begin{bmatrix} a_1 \\ a_2 \end{bmatrix}=s_+\begin{bmatrix} a_1 \\ a_2 \end{bmatrix} \tag{9.6}$$

Hence

$$\frac{a_1}{a_2}=j \tag{9.7}$$

A normalized eigenvector is, therefore,

$$\bar{U}_+=\frac{1}{\sqrt{2}}\begin{bmatrix} 1 \\ -j \end{bmatrix} \tag{9.8}$$

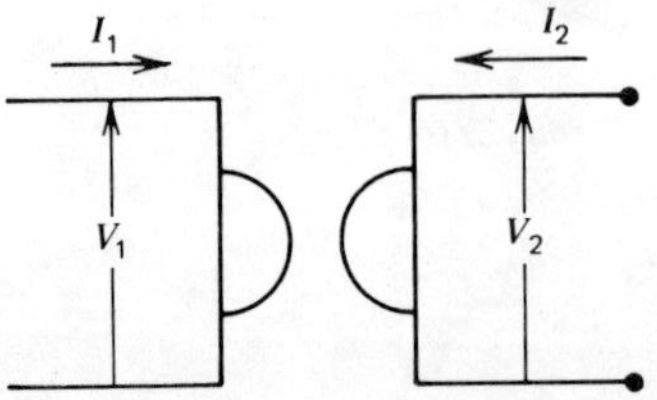

Figure 9.4. Schematic of gyrator network.

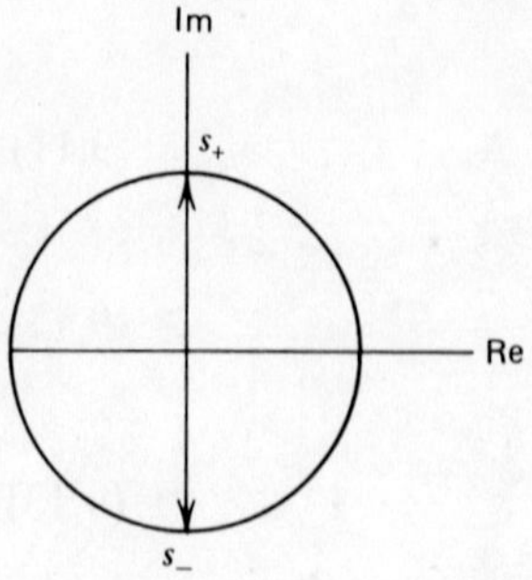

Figure 9.5. Eigenvalues of gyrator network.

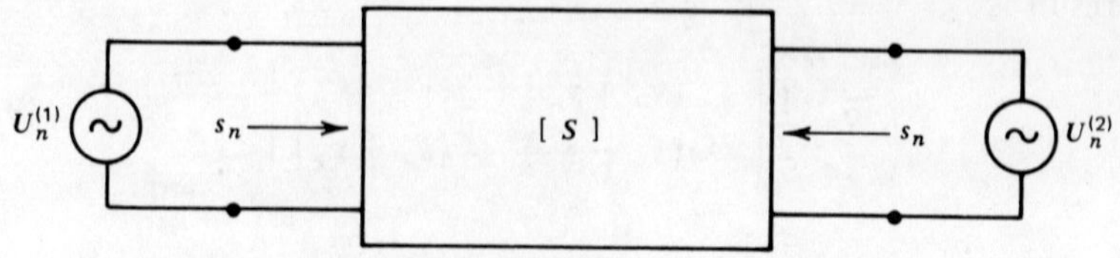

Figure 9.6a. Schematic diagram of eigenvalue equation.

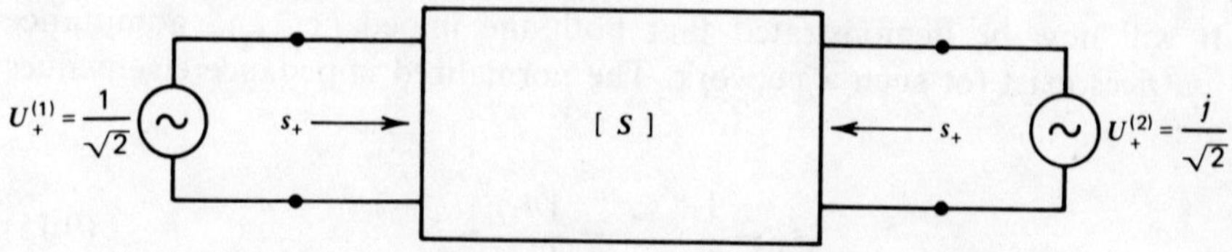

Figure 9.6b. Schematic diagram of U_+ eigenvalue equation.

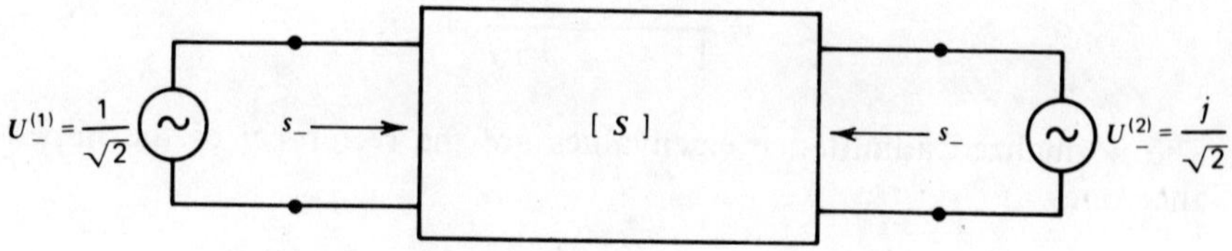

Figure 9.6c. Schematic diagram of U_- eigenvalue equation.

In a similar way one has

$$\overline{U}_- = \frac{1}{\sqrt{2}}\begin{bmatrix} 1 \\ j \end{bmatrix} \tag{9.9}$$

Figures 9.6*b* and 9.6*c* define Eq. 9.5 for the two eigenvectors.

The scattering matrix is obtained by diagonalizing the matrix $\overline{S}$.

$$\overline{S} = \overline{U}\,\overline{\lambda}(\overline{U}^*)^T \tag{9.10}$$

where

$$\bar{U} = \frac{1}{\sqrt{2}} \begin{bmatrix} 1 & 1 \\ -j & j \end{bmatrix} \tag{9.11}$$

$$(\bar{U}^*)^T = \frac{1}{\sqrt{2}} \begin{bmatrix} 1 & j \\ 1 & -j \end{bmatrix} \tag{9.12}$$

$$\bar{\lambda} = \begin{bmatrix} s_+ & 0 \\ 0 & s_- \end{bmatrix} \tag{9.13}$$

The result is

$$\bar{S} = \frac{1}{2} \begin{bmatrix} (s_+ + s_-) & j(s_+ - s_-) \\ -j(s_+ - s_-) & (s_+ + s_-) \end{bmatrix} \tag{9.14}$$

9.2 *IMMITTANCE MATRICES OF GYRATOR NETWORK*

It will now be demonstrated that both the impedance and admittance matrices exist for such a network. The normalized impedance eigenvalues are

$$z_+ = \frac{1+s_+}{1-s_+} = \frac{1+j}{1-j} = j \tag{9.15}$$

$$z_- = \frac{1+s_-}{1-s_-} = \frac{1-j}{1+j} = -j \tag{9.16}$$

The normalized admittance eigenvalues are the reciprocal of the impedance ones.

$$y_+ = \frac{1}{z_+} = -j \tag{9.17}$$

$$y_- = \frac{1}{z_-} = j \tag{9.18}$$

The admittance matrix is now diagonalized using Eq. 9.10

$$\bar{Y} = \frac{1}{2} \begin{bmatrix} (y_+ + y_-) & j(y_+ - y_-) \\ -j(y_+ - y_-) & (y_+ + y_-) \end{bmatrix} \tag{9.19}$$

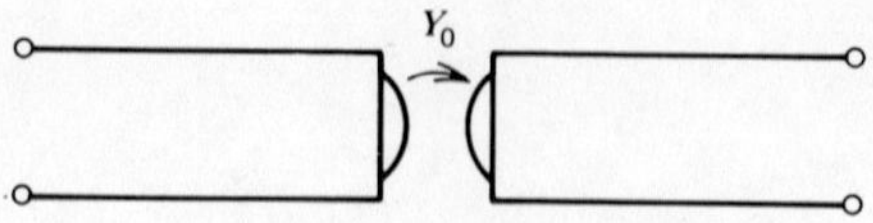

Figure 9.7. $\bar{Y}$ matrix equivalent circuit.

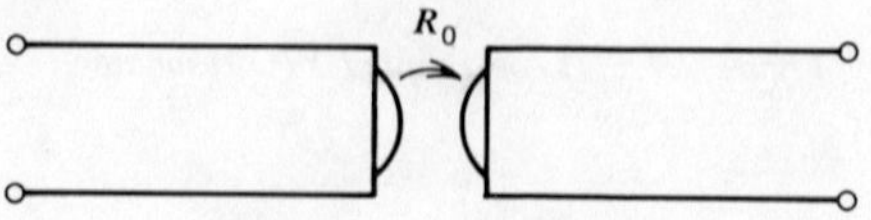

Figure 9.8. $\bar{Z}$ matrix equivalent circuit.

In terms of the original variables the result is

$$\bar{Y}=\begin{bmatrix} 0 & 1 \\ -1 & 0 \end{bmatrix} \tag{9.20}$$

The $\bar{Y}$-matrix schematic of the 2-port gyrator is illustrated in Figure 9.7.

The result for the $\bar{Z}$-matrix is

$$\bar{Z}=\begin{bmatrix} 0 & -1 \\ 1 & 0 \end{bmatrix} \tag{9.21}$$

The $\bar{Z}$-matrix schematic of the 2-port gyrator is shown in Figure 9.8.

9.3 *TWO-PORT GYRATOR USING ORTHOGONAL LOOPS COUPLED BY A YIG SPHERE*

There are a number of different physical ways in which gyrator networks can be constructed. One circuit arrangement, which contains a 2-port gyrator, is that of the bandpass filter in Figure 9.9*a*. Two coils have their axes at right angles to each other, and a small ferrite sample is placed at the intersection of the coil axes. When the sample is not magnetized, no power is transferred between the coils because the loop axes are perpendicular to each other and there is no interaction with the ferrite. When a direct field is applied along the z axis, the two coils are coupled through the transverse components of the dipolar field of the ferrite resonator. This coupling is largest at ferrimagnetic resonance.

The $\bar{Z}$ matrix of the loop coupled YIG resonator may be obtained directly from its schematic diagram in Figure 9.9*a* but in keeping with the eigenvalue approach used in this text it will be obtained from its eigennetworks in Figures 9.9*b* and 9.9*c*. Since the two loops are decoupled, the eigennetworks and the original circuit have common loop geometries. However, the tensor form of the susceptibility in the original circuit takes on the scalar values $\chi_{\pm}$ in Eqs. 8.30 and 8.31 for the two eigennetworks.

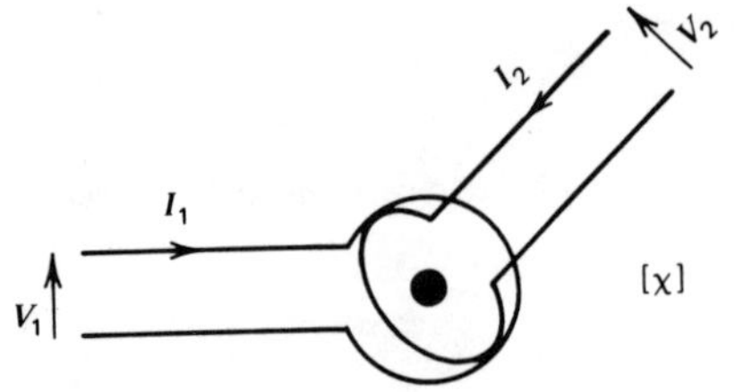

Figure 9.9a. Loop-coupled YIG resonator.

Figure 9.9b. $\overline{U}_+$ *eigennetwork of loop coupled YIG resonator.*

Figure 9.9c. $\overline{U}_-$ *eigennetwork of loop coupled YIG resonator.*

The derivation of the impedance eigenvalues starts by assuming counter-rotating magnetic fields $h_\pm$ at the ferrite sample due to currents $I_\pm$ in the two eigennetworks.

$$h_+^e = KI_+ \tag{9.22}$$

$$h_-^e = KI_- \tag{9.23}$$

where the coupling coefficient K depends only on the geometry of the circuit and the position of the ferrite sample.

The voltages induced in the two eigennetworks are given by

$$V_+ = j\omega \int_{\text{loop}} b_+ \, da \tag{9.24}$$

$$V_- = j\omega \int_{\text{loop}} b_- \, da \tag{9.25}$$

The inductions at the loops due to the equivalent magnetic dipolar fields of the ferrite resonator are

$$b_+ = \frac{m_+ v}{2\pi r^3} \tag{9.26}$$

$$b_- = \frac{m_- v}{2\pi r^3} \tag{9.27}$$

where v is the volume of the ferrite resonator and r is the radial distance from the magnetic dipole.

The transverse magnetization in the sample is given

$$m_+ = \mu_0 \chi_+^e h_+^e \tag{9.28}$$

$$m_- = \mu_0 \chi_-^e h_-^e \tag{9.29}$$

where $\chi_\pm^e$ is given in chapter 8 by

$$\chi_\pm^e = \frac{\omega_m}{\mp\omega + \omega_r} \tag{9.30}$$

If we substitute for m_+ and m_- in Eqs. 9.26 and 9.27

$$b_+ = \frac{\mu_0 v}{2\pi r^3}(\chi_+^e K I_+) \tag{9.31}$$

$$b_- = \frac{\mu_0 v}{2\pi r^3}(\chi_-^e K I_-) \tag{9.32}$$

Finally, introducing the above equations into Eqs. 9.24 and 9.25 gives the impedance eigenvalues as

$$Z_+ = \frac{j\mu_0 \omega v}{2\pi} \int_{\text{loop}} \frac{\chi_+^e K}{r^3} da \tag{9.33}$$

$$Z_- = \frac{j\mu_0 \omega v}{2\pi} \int_{\text{loop}} \frac{\chi_-^e K}{r^3} da \tag{9.34}$$

As an example of the use of this result, consider the coupling to two orthogonal loops (radius r_0) by a ferrite resonator in the geometrical center as shown in Figure 9.2. In this case

$$h_+^e = \frac{I_+}{2r_0} \tag{9.35}$$

$$h_-^e = \frac{I_-}{2r_0} \tag{9.36}$$

This gives

$$K = \frac{1}{2r_0} \tag{9.37}$$

in Eqs. 9.33 and 9.34.

When integrated over the area of the loops, Eqs. 9.33 and 9.34 become

$$Z_+ = \frac{j\mu_0 \omega v \chi_+^e}{2r_0^2} \tag{9.38}$$

$$Z_- = \frac{j\mu_0 \omega v \chi_-^e}{2r_0^2} \tag{9.39}$$

Taking linear combinations of $Z_\pm$ gives the standard result for the $\overline{Z}$ matrix

$$\begin{bmatrix} V_1 \\ V_2 \end{bmatrix} = \begin{bmatrix} Z_{11} & -R_0 \\ R_0 & Z_{11} \end{bmatrix} \begin{bmatrix} I_1 \\ I_2 \end{bmatrix} \tag{9.40}$$

V_1 and V_2 are the voltages developed around the loops and I_1 and I_2 are the currents flowing in the loops.

In the preceding equation,

$$Z_{11} = \frac{j\mu_0 \omega v \chi_{xx}^e}{2r_0^2} \tag{9.41}$$

$$R_0 = \frac{-j\mu_0 \omega v \chi_{xy}^e}{2r_0^2} \tag{9.42}$$

and

$$\chi_{xx}^e = \frac{\chi_+^e + \chi_-^e}{2} \tag{9.43}$$

$$\chi_{xy}^e = \frac{\chi_+^e - \chi_-^e}{2} \tag{9.44}$$

Equations 9.38 and 9.39 may be modified to account for the self-inductance L_s of the two loops as follows

$$Z_+ = j\omega L_s + \frac{j\mu_0 \omega v \chi_+^e}{2r_0^2} \tag{9.45}$$

$$Z_- = j\omega L_s + \frac{j\mu_0 \omega v \chi_-^e}{2r_0^2} \tag{9.46}$$

9.4 EQUIVALENT CIRCUIT OF COUPLED YIG RESONATOR USING GYRATOR NETWORK

One exact symmetrical equivalent circuit for the coupled YIG resonator based on an ideal gyrator circuit is due to Carter.[2] This circuit is indicated in Figure 9.10*a*. The shunt inductance L_f is a ferrite induced one while the series inductance L_s is the self-inductance of the coupling networks. Figures 9.10*b* and 9.10*c* depict the two eigennetworks for this circuit in terms of the impedance eigenvalues $Z'_{\pm}$ of the ideal gyrator network.

The equivalence between the eigennetworks in Figures 9.10*b* and 9.10*c* and those in Figures 9.9*b* and 9.9*c* for the coupled YIG resonator may be obtained by comparing the two descriptions. For the impedance matrix description of the resonator the two eigenvalues are given by combining Eqs. 9.45 and 9.46 with Eq. 9.30

$$Z_+ = j\omega L_s + \frac{j\mu_0 \omega v \omega_m}{(-\omega+\omega_r)2r_0^2} \tag{9.47}$$

$$Z_- = j\omega L_s + \frac{j\mu_0 \omega v \omega_m}{(\omega+\omega_r)2r_0^2} \tag{9.48}$$

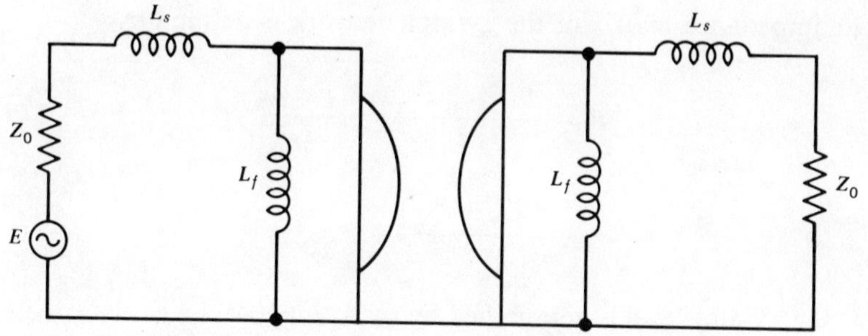

Figure 9.10a. Exact equivalent circuit of coupled YIG resonator.

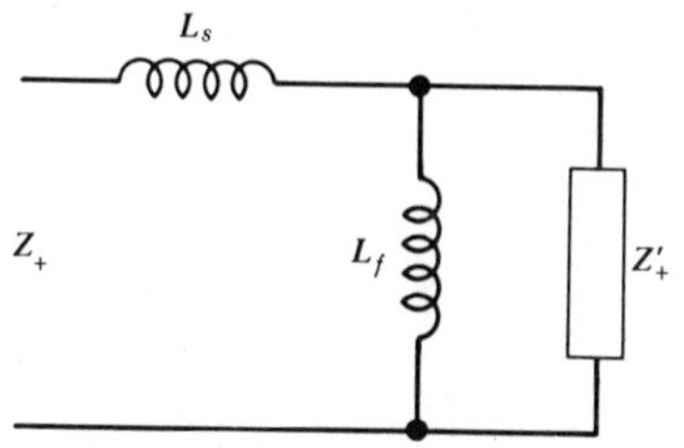

Figure 9.10b. $\overline{U}_+$ eigennetwork of equivalent circuit of coupled YIG resonator.

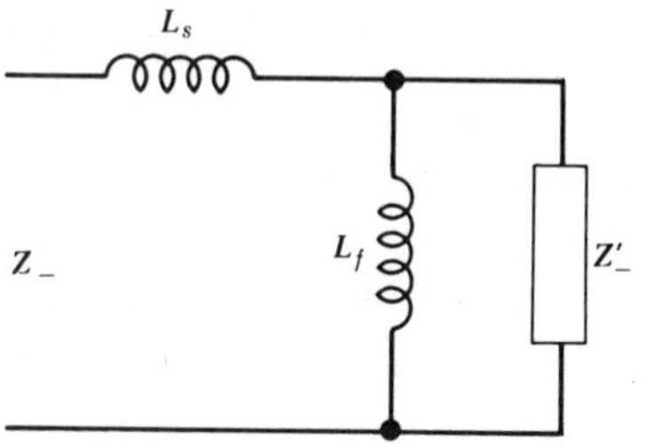

Figure 9.10c. $\overline{U}_-$ eigennetwork of equivalent circuit of coupled YIG resonator.

For the eigennetworks in Figures 9.10*b* and 9.10*c* the result is

$$Z_{+}=j\omega L_s+\frac{j\omega L_f(Z'_{+}/-jL_f)}{-\omega+(Z'_{+}/-jL_f)} \tag{9.49}$$

$$Z_{-1}=j\omega L_s+\frac{j\omega L_f(Z'_{-}/jL_f)}{\omega+(Z'_{-}/jL_f)} \tag{9.50}$$

The two sets of eigenvalues are equivalent provided

$$Z'_{+}=-j\omega_r L_f \tag{9.51}$$

$$Z'_{-}=j\omega_r L_f \tag{9.52}$$

where

$$L_f=\frac{\mu_0 v\omega_m}{2r_0^2\omega_r} \tag{9.53}$$

Thus the eigenvalues of the gyrator network have the nature defined by Eqs. 9.15 and 9.16.

The impedance matrix of the gyrator network is defined by

$$R_0=\frac{j(Z'_{+}-Z'_{-})}{2}=\frac{\mu_0 v\omega_m}{2r_0^2} \tag{9.54}$$

$$Z_{11}=\frac{(Z'_{+}+Z'_{-})}{2}=0 \tag{9.55}$$

Thus the gyrator level is determined by the volume of the garnet sphere, by the radius of the coupling loop and by the magnetization of the garnet material.

9.5 *SCATTERING MATRIX OF LOOP COUPLED YIG FILTER AT RESONANCE*

The impedance eigenvalues for the YIG filter are

$$Z_{+}=\frac{j\omega R_0}{-\omega+\omega_r} \tag{9.56}$$

$$Z_{-}=\frac{j\omega R_0}{\omega+\omega_r} \tag{9.57}$$

Forming the corresponding scattering eigenvalues through the 1-port relation between s and Z gives

$$s_+ = \frac{(j\omega R_0/Z_0) - (-\omega+\omega_r)}{(j\omega R_0/Z_0) + (-\omega+\omega_r)} \tag{9.58}$$

$$s_- = \frac{(j\omega R_0/Z_0) - (\omega+\omega_r)}{(j\omega R_0/Z_0) + (\omega+\omega_r)} \tag{9.59}$$

At resonance $s_\pm$ are

$$s_+ = 1 \tag{9.60}$$

$$s_- = \frac{(jR_0/Z_0) - 2}{(jR_0/Z_0) + 2} \tag{9.61}$$

The scattering parameters are therefore given at resonance by

$$|S(0)_{21}|^2 = \frac{1}{1 + (R_0/2Z_0)^2} \tag{9.62}$$

$$|S(0)_{11}|^2 = \frac{(R_0/2Z_0)^2}{1 + (R_0/2Z_0)^2} \tag{9.63}$$

9.6 *BANDWIDTH OF LOOP-COUPLED YIG FILTER*

In the vicinity of the main resonance, Eqs. 9.58 and 9.59 become

$$s_+ \approx \frac{(jR_0/Z_0) - \delta}{(jR_0/Z_0) + \delta} \tag{9.64}$$

$$s_- \approx \frac{(jR_0/Z_0) - 2}{(jR_0/Z_0) + 2} \tag{9.65}$$

where

$$\delta \approx \frac{-\omega+\omega_r}{\omega} \tag{9.66}$$

Writing S_{21} in terms of the original variables gives

$$|S_{21}|^2 = \frac{\left[(\delta R_0/Z_0) - (2R_0/Z_0)\right]^2}{\left[-(R_0/Z_0)^2 + 2\delta\right]^2 + \left[(2R_0/Z_0) + (\delta R_0/Z_0)\right]^2} \tag{9.67}$$

For $\delta \ll 2$

$$|S_{21}|^2 \approx \frac{|S(0)_{21}|^2}{1+|S(0)_{21}|^2(Z_0/2R_0)^2(2\delta)^2} \tag{9.68}$$

where $|S(0)_{21}|^2$ is defined by Eq. 9.62

The bandwidth of the device at the 3-dB points is now given by

$$2\delta = \frac{1}{|S(0)_{21}|(Z_0/2R_0)} \tag{9.69}$$

9.7 *SCATTERING MATRIX OF LOOP-COUPLED YIG FILTER WITH DAMPING*

Introducing damping into Eqs. 9.56 and 9.57 by replacing ω_r by $\omega_r + j\alpha\omega$ in Eq. 9.30 gives

$$Z_+ = \frac{j\omega R_0}{-\omega+\omega_r+j\alpha\omega} \tag{9.70}$$

$$Z_- = \frac{j\omega R_0}{\omega+\omega_r+j\alpha\omega} \tag{9.71}$$

where

$$\alpha = \frac{\gamma \Delta H}{2\omega} \tag{9.72}$$

ΔH is the uniform mode line width and γ is the gyromagnetic ratio. Neglecting the effect of damping on the eigennetwork which does not exhibit the main resonance, the preceding equations become at resonance

$$Z_+ = \frac{R_0}{\alpha} \tag{9.73}$$

$$Z_- \approx \frac{jR_0}{2} \tag{9.74}$$

Thus

$$s_+ = \frac{(R_0/\alpha Z_0)-1}{(R_0/\alpha Z_0)+1} \tag{9.75}$$

$$s_- \approx \frac{(jR_0/2Z_0)-1}{(jR_0/2Z_0)+1} \tag{9.76}$$

Constructing S_{21} in the usual way in terms of $s_\pm$ gives the required result

$$|S_{21}|^2 \approx \frac{|S(0)_{21}|^2}{[1+(\alpha Z_0/R_0)]^2} \tag{9.77}$$

where $|S(0)_{21}|$ is the scattering coefficient in the absence of damping defined by Eq. 9.62, and it has been assumed that $\alpha \ll 2$.

The dissipation in the circuit is also readily obtained by constructing the dissipation in each eigennetwork

$$q_+ = 1 - s_+ s_+^* \approx \frac{4\alpha Z_0}{R_0} \tag{9.78}$$

$$q_- = 1 - s_- s_-^* = 0 \tag{9.79}$$

Thus the circuit dissipation is given by

$$Q_{11} = \frac{q_+ + q_-}{2} \approx \frac{2\alpha Z_0}{R_0} \tag{9.80}$$

$|S_{21}|^2$ may now also be written in terms of the dissipation coefficient Q_{11} as

$$|S_{21}|^2 \approx |S(0)_{21}|^2 (1 - Q_{11}) \tag{9.81}$$

Finally, S_{11} is now determined from the energy condition

$$1 - |S_{11}|^2 - |S_{21}|^2 = Q_{11} \tag{9.82}$$

or by directly forming S_{11} in terms of $s_\pm$. The result is

$$|S_{11}|^2 \approx 1 - \frac{2\alpha Z_0}{R_0} - \frac{|S(0)_{21}|^2}{[1+(\alpha Z_0/R_0)]^2} \tag{9.83}$$

9.8 *REFLECTION COEFFICIENT OF LOSSY GYRATOR CIRCUIT WITH ONE PORT TERMINATED IN VARIABLE SHORTCIRCUIT*

One way the damping in the two eigennetworks of the gyrator circuit may be separately evaluated is by measuring the reflection coefficient at its input terminals with the output ones terminated in a variable shortcircuit. The input-output relation of the gyrator network is

$$\begin{bmatrix} b_1 \\ b_2 \end{bmatrix} = \begin{bmatrix} S_{11} & S_{12} \\ S_{21} & S_{11} \end{bmatrix} \begin{bmatrix} a_1 \\ a_2 \end{bmatrix} \tag{9.84}$$

Letting

$$a_1 = 1 \tag{9.85}$$

$$a_2 = \exp(-j\phi) \tag{9.86}$$

gives

$$\frac{b_1}{a_1} = S'_{11} = S_{11} + \frac{S_{12}S_{21}\exp(-j\phi)}{1 - S_{11}\exp(-j\phi)} \tag{9.87}$$

The eigenvalues of the scattering matrix for an ideal gyrator with damping are

$$s_+ \approx j\left(1 - \frac{q_+}{2}\right) \tag{9.88}$$

$$s_- \approx -j\left(1 - \frac{q_-}{2}\right) \tag{9.89}$$

where $q_\pm$ are the eigenvalues of the dissipation matrix or equivalently, the dissipation in each eigennetwork.

Writing S_{11}, S_{12}, and S_{21} in terms of $s_\pm$ in the usual way gives

$$S'_{11} \approx j(1 - q_+) \tag{9.90}$$

for $\exp(-j\phi) = j$, and

$$S'_{11} \approx -j(1 - q_-) \tag{9.91}$$

for $\exp(-j\phi) = -j$.

For the intermediate positions of the variable shortcircuit (where $\exp(-j\phi) = \pm 1$) the result is

$$S'_{11} = \pm\left(1 - \frac{q_+ + q_-}{2}\right) \tag{9.92}$$

Thus Eqs. 9.90 and 9.91 represent the two extreme values of S'_{11} and may therefore be used to determine the damping parameters $q_\pm$ of the eigennetworks and thereby the overall loss of the circuit.

The dissipation eigenvalues may be readily measured by connecting the circuit between the output terminals of a 3-dB hybrid in the manner indicated in Figures 9.11*a* and 9.11*b*. This arrangement also leads to a tunable resonance isolator-type structure between the input terminals of the hybrid.

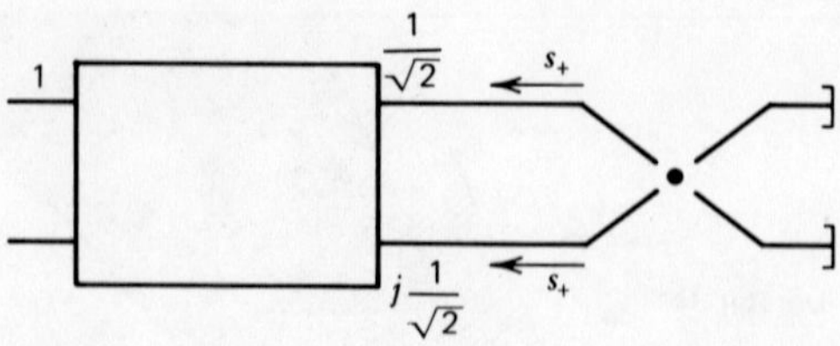

Figure 9.11a. *Circuit arrangement for measuring s_+ eigenvalue of gyrator circuit.*

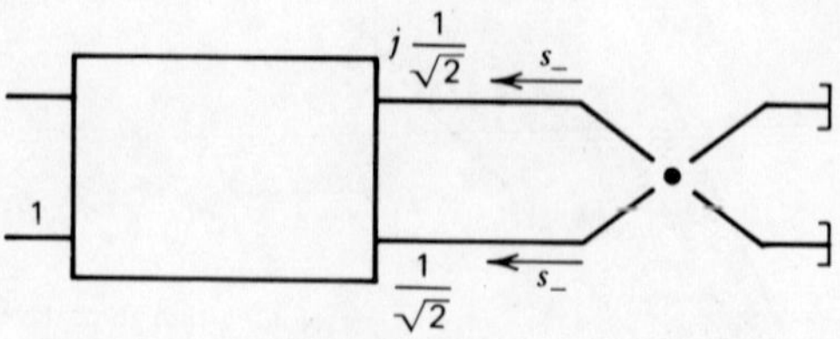

Figure 9.11b. *Circuit arrangement for measuring s_- eigenvalue of gyrator circuit.*

9.9 *COINCIDENCE LIMITING IN YIG SPHERES AT LARGE SIGNAL POWER*

YIG resonators exhibit nonlinear attenuation at large signal levels due to the coupling of the uniform mode to spinwaves (higher order modes of the magnetization) provided certain frequency relations are satisfied. The most important instability at the main ferrimagnetic resonance is known as coincidence limiting. In this arrangement the microwave frequency coincides both with the main resonance and with spinwaves at half the pump frequency. The effect of such nonlinear loss is to degrade the unloaded Q factor of the YIG resonator.

For the coincidence condition to hold it is necessary to satisfy the following expression:

$$\omega > 2N_t(\gamma 4\pi M_0)$$

This is an essential requirement for the coincidence condition. Since the sample must be magnetically saturated it is also necessary to have

$$N_t(\gamma 4\pi M_0) > \omega$$

Combining the above two conditions gives the coincidence limiting interval as

$$N_t(\gamma 4\pi M_0) > \omega > 2N_t(\gamma 4\pi M_0)$$

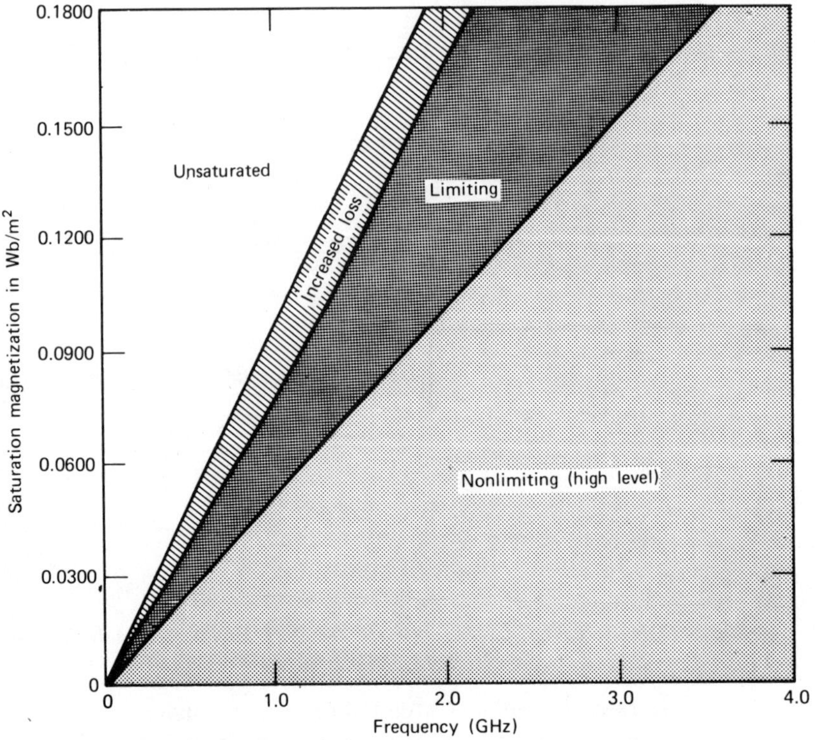

Figure 9.12. Coincidence limiting region of YIG resonator.

where N_t is the transverse demagnetizing factor equal to 1/3 for a sphere. Figure 9.12 depicts the unsaturated, coincidence, and nonlimiting regions for a sphere as a function of saturation magnetization.

PROBLEMS

1. Starting with Eqs. 9.78 and 9.79 obtain the $\overline{Q}$ matrix of the loop-coupled YIG filter.
2. Obtain S_{11} and S_{21} for the loop-coupled YIG filter by retaining the self-inductance of the loops L_s in Eqs. 9.56 and 9.57.
3. Construct S_{11} described by Eq. 9.83 using the eigenvalue approach.
4. Discuss the power dissipated in the YIG filter for different incident waves with the help of $P_{\text{diss}} = \frac{1}{2}\tilde{a}^* \overline{Q}\bar{a}$.
5. Obtain the $\overline{Z}$ matrix of the loop-coupled YIG filter given by Eq. 9.40 using the transverse form for the susceptibility defined in Chapter 8.

6. Using Eq. 9.54 calculate R_0 at 10 GHz for a garnet sphere 1 mm in diameter with a magnetization of 0.1780 Wb/m^2 and a loop diameter $2r_0$ of 1.2 mm.
7. Obtain $S(0)_{21}$ and $S(0)_{11}$ for the sphere in problem 6.
8. Calculate the 3-dB B.W. of a YIG filter using the sphere in problem 6.
9. A YIG filter has a sphere with a magnetization of 0.1780 Wb/m^2 and a demagnetizing factor N_t of 1/3. Find its tuning range.
10. Obtain the input impedance of an ideal gyrator circuit terminated by an impedance Z_L. Show that such a circuit exhibits an inductive reactance at its input terminals when terminated by a capacitive one.

FURTHER READING

1. R. W. DeGrasse, Low-loss gyromagnetic coupling through single crystal garnets, *J. Appl. Phys.*, **30S** (April 1959), pp. 155S–156S.
2. P. S. Carter, Equivalent circuit of orthogonal-loop-coupled magnetic resonance filters and bandwidth narrowing due to coupling inductance, *IEEE Trans. Microwave Theory Tech.*, **18** (Feb. 1970), pp. 100–105.
3. Y. Konishi and Y. Utsumi, Ferrite tuner with polycrystalline Ca V garnet, *IEEE Trans. Magn.* (1972 INTERMAG Conf.), (Sept. 1972), pp. 547–549.
4. H. Bex, Theorie des Magnetisch Abstimmbaren Bandpass-filters, *Nachrichtentech. Z.*, (Sept. 1972), pp. 25–29. and pp. 390–394.
5. H. Skeie, Nonreciprocal coupling with single-crystal ferrites, *IEEE Trans. Microwave Theory Tech.*, **12**, (Nov. 1964), pp. 587–594.
6. M. Igarashi and Y. Naito, Properties of a four-port non-reciprocal circuit utilizing YIG on stripline—Filter and circulator, *IEEE Trans. Microwave Theory Tech.*, **20**, (Dec. 1972), pp. 828–833.
7. M. Igarashi and Y. Naito, Theoretical Analysis of Magnetic Resonance Nonreciprocal Circuits—Limitations of 3-dB Bandwidth and Available Range, *IEEE Trans. Microwave Theory Tech.*, **22**, 9 (Sept. 1974).
8. J. Helszajn, *Nonreciprocal Microwave Junctions and Circulators*, Wiley, New York, 1975.
9. R. L. Comstock, Synthesis of filter-limiters using ferrimagnetic resonators, *IEEE Trans. Microwave Theory Tech.*, (Nov. 1964).

REFERENCES

1. Tellegen "The Microwave Gyrator," *Bell Syst. Tech. J.* (1952).
2. P. S. Carter, Jr., Magnetically-tunable microwave filters using single-crystal yttrium-iron-garnet resonators, *IRE Trans. Microwave Theory Tech.*, **9** (May 1961), pp. 252–260.

CHAPTER TEN

The Junction Circulator

The most general form of the 3-port junction circulator consists of a symmetrical distribution of ferrite material at the junction of three transmission lines. One stripline geometry is shown in Figures 10.1*a* and 10.1*b*. It consists of two ferrite disk resonators separated by a disk center conductor fed by symmetrical transmission lines. The ferrite material is magnetized perpendicularly to the plane of the disk by a static magnetic field. In this device power entering port 1 emerges from port 2, and so on, in a cyclic manner. An important property of the device is that a perfect circulator is obtained when the junction is matched. For a three-port junction this requires two independent variables. Under certain simplifying

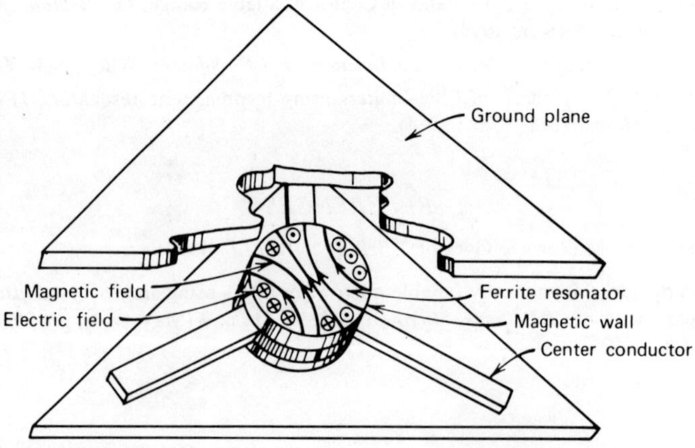

Figure 10.1a. Three-port stripline junction showing field patterns in demagnetized ferrite resonator.

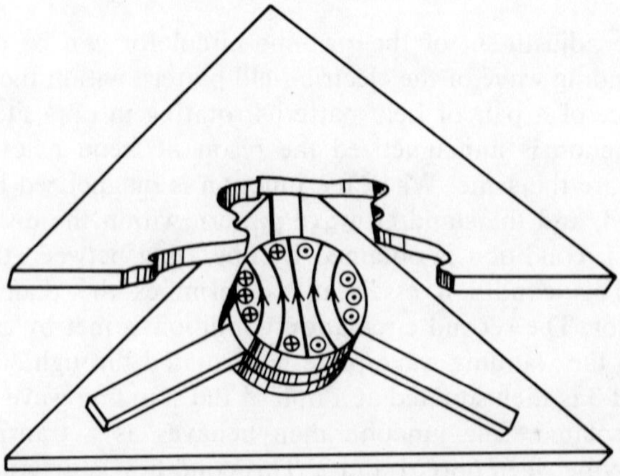

Figure 10.1b. Three-port stripline circulator showing field patterns in magnetized ferrite resonator.

Figure 10.2. Schematic of commercial waveguide circulator (courtesy Ferranti Ltd.).

conditions the adjustment of the stripline circulator can be described in terms of a standing wave of the electric-field pattern within the disk due to the interference of a pair of field patterns rotating in opposite directions. When the junction is unmagnetized the resonant frequencies of the two field patterns are the same. When the junction is magnetized the degeneracy is removed, and the standing wave pattern within the disk is rotated. One circulation condition is obtained by operating between the two split frequencies. This condition essentially determines the diameter of the ferrite resonator. The second circulation condition is met by adjusting the splitting, until the standing wave pattern is rotated through 30 deg. From symmetry port 3 is then situated at a null of the standing wave pattern and is therefore isolated. The junction then behaves as a transmission line resonant cavity between ports 1 and 2. This condition is determined by the magnitude of the direct magnetic field.

Figure 10.1*a* gives the field pattern in a demagnetized 3-port junction and Figure 10.1*b* depicts the same field pattern rotated through 30 deg to form an ideal circulator. Figure 10.2 shows a commercial version of a waveguide circulator.

10.1 *NETWORK DEFINITION OF JUNCTION CIRCULATOR*

Many of the properties of junction circulators are best discussed in terms of the scattering matrix of the device defined in Figure 10.3. For a lossless junction the unitary properties of the scattering matrix may be invoked to define the ideal 3-port circulator discussed in this text.

For such a junction the principle of the conservation of energy requires that the scattering matrix be unitary

$$\bar{S}\,(\bar{S}^*)^T = \bar{I} \tag{10.1}$$

It will now be demonstrated that a matched symmetric 3-port junction is necessarily a circulator. The scattering matrix of this junction is given by

$$\bar{S} = \begin{bmatrix} 0 & S_{12} & S_{13} \\ S_{13} & 0 & S_{12} \\ S_{12} & S_{13} & 0 \end{bmatrix} \tag{10.2}$$

Applying the unitary condition to this matrix gives

$$|S_{12}|^2 + |S_{13}|^2 = 1 \tag{10.3}$$

and

$$S_{12}S_{13}^* = 0 \tag{10.4}$$

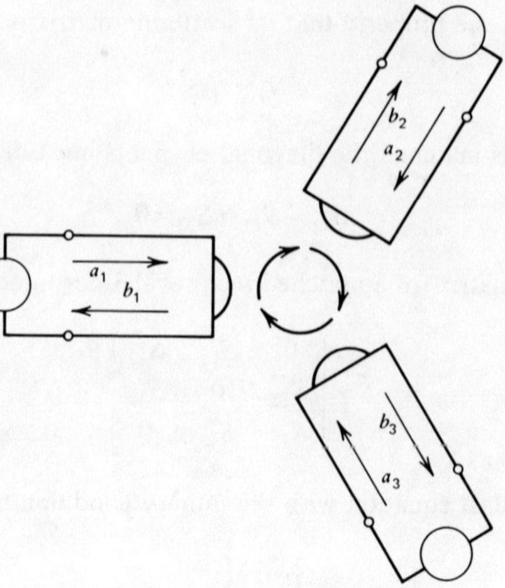

$$\begin{bmatrix} b_1 \\ b_2 \\ b_3 \end{bmatrix} = \begin{bmatrix} S_{11} & S_{12} & S_{13} \\ S_{13} & S_{11} & S_{12} \\ S_{12} & S_{13} & S_{11} \end{bmatrix} \begin{bmatrix} a_1 \\ a_2 \\ a_3 \end{bmatrix}$$

Figure 10.3. Schematic diagram of 3-port junction showing input and output scattering waves.

The last two equations are consistent provided

$$|S_{12}|=1 \tag{10.5}$$

and

$$|S_{13}|=0 \tag{10.6}$$

or

$$|S_{12}|=0 \tag{10.7}$$

$$|S_{13}|=1 \tag{10.8}$$

It is obvious from the above relations that if $S_{11}=0$ (matched condition), then either $|S_{12}|=0$ and $|S_{13}|=1$, or $|S_{13}|=0$ and $|S_{12}|=1$. These are the conditions for a perfect circulator.

The unitary condition will now be used to demonstrate that it is not possible to match a lossless reciprocal 3-port junction. If the junction is reciprocal it has the property that its scattering matrix is symmetric

$$S_{ij} = S_{ji} \tag{10.9}$$

If the junction is matched the diagonal elements must be zero

$$S_{11} = S_{22} = S_{33} = 0 \tag{10.10}$$

The scattering matrix for a matched reciprocal 3-port junction is, therefore,

$$\bar{S} = \begin{bmatrix} 0 & S_{12} & S_{12} \\ S_{12} & 0 & S_{12} \\ S_{12} & S_{12} & 0 \end{bmatrix} \tag{10.11}$$

Combining the last equation with the unitary condition gives

$$2S_{12}S_{12}^{*} = 1 \tag{10.12}$$

and

$$S_{12}S_{12}^{*} = 0 \tag{10.13}$$

It is therefore impossible to match a reciprocal lossless 3-port junction, since it is not possible to satisfy Eqs. 10.12 and 10.13 simultaneously. It will be shown later on that the best possible match for a reciprocal 3-port junction occurs when $S_{11} = 1/3$.

Although the unitary condition may be used to define an ideal circulator, it does not indicate in what way such a network may be realized. To understand the operation of the junction circulator, it is necessary to establish the eigenvalues s_n and eigenvectors $\bar{U}_n$ associated with the matrix $\bar{S}$.

10.2 *THE SCATTERING MATRIX EIGENVALUES*

The relation between the scattering matrix and its eigenvalues can be obtained from the eigenvalue equation of the square matrix $\bar{S}$

$$\bar{S}\,\bar{U}_n = s_n \bar{U}_n \tag{10.14}$$

where $\bar{U}_n$ is an eigenvector and s_n is an eigenvalue. This equation is illustrated in Figure 10.4.

Equation 10.14 has a nonvanishing value for U_n provided

$$\det|\bar{S} - s_n \bar{I}| = 0 \tag{10.15}$$

where $\bar{I}$ is a unit matrix.

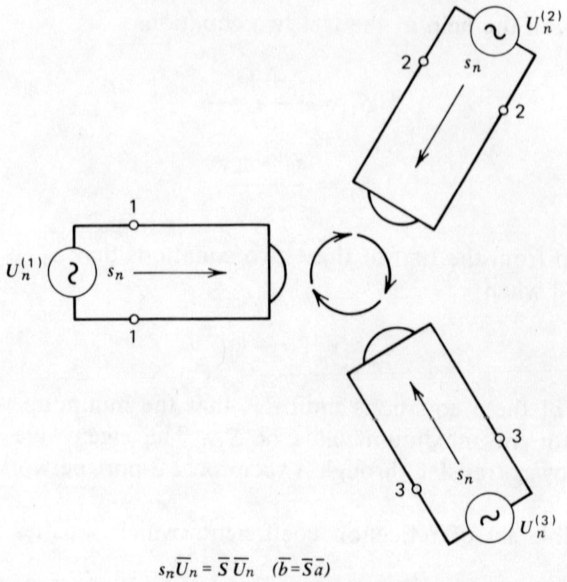

Figure 10.4. Schematic diagram of 3-port junction indicating definition of eigenvalue equation.

Since the eigenvalues of a reciprocal 3-port junction for which S_{11} is a minimum are related to those of an ideal 3-port circulator they will be derived first: For a reciprocal 3-port junction the $\bar{S}$ matrix is

$$\bar{S} = \begin{bmatrix} S_{11} & S_{12} & S_{12} \\ S_{12} & S_{11} & S_{12} \\ S_{12} & S_{12} & S_{11} \end{bmatrix} \tag{10.16}$$

The characteristic equation for this matrix is

$$(S_{11} - s_n)^3 - 3(S_{11} - s_n)S_{12}^2 + 2S_{12}^3 = 0 \tag{10.17}$$

The three eigenvalues are

$$s_0 = S_{11} + 2S_{12} \tag{10.18}$$

$$s_{+1} = s_{-1} = S_{11} - S_{12} \tag{10.19}$$

This result indicates that two of the eigenvalues in a reciprocal 3-port junction are degenerate.

The entries of the $\bar{S}$ matrix can also be written in terms of the eigenvalues with the help of the last two equations

$$S_{11}=\frac{s_0+2s_{+1}}{3} \tag{10.20}$$

$$S_{12}=\frac{s_0-s_{+1}}{3} \tag{10.21}$$

It is observed from the first of these two equations that S_{11} is a minimum equal to $|1/3|$ when

$$s_{+1}=-s_0 \tag{10.22}$$

The second of these equations indicates that the minimum value for S_{11} coincides with the maximum value of S_{12}. The eigenvalue diagram for maximum power transfer through a reciprocal 3-port network is depicted in Figure 10.5.

One possible set of reflection coefficients which satisfies the last two equations is

$$s_{+1}=e^{-j2(\theta_1+\pi/2)} \tag{10.23}$$

$$s_{-1}=e^{-j2(\theta_1+\pi/2)} \tag{10.24}$$

$$s_0=e^{-j2\theta_0} \tag{10.25}$$

provided

$$\theta_0=\theta_1=\frac{\pi}{2} \tag{10.26}$$

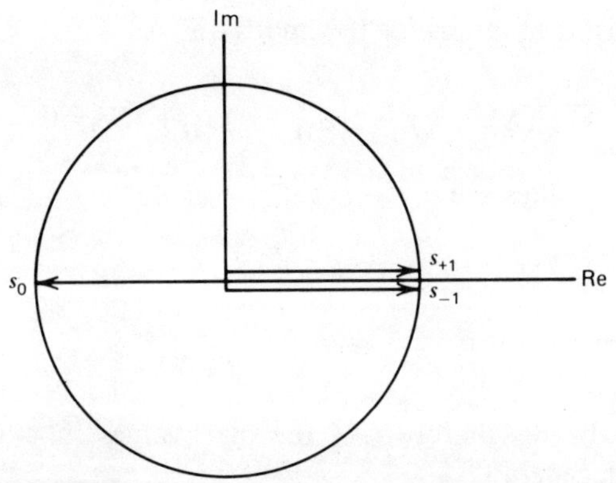

Figure 10.5. Eigenvalue diagram of reciprocal 3-port junction for maximum power transfer.

These reflection coefficients are associated with the 1-port reactive transmission lines indicated in Figure 10.6. The origin of the opencircuited and shortcircuited terminations will be discussed in connection with the eigenvectors of the junction.

The scattering matrix for this arrangement is

$$\bar{S}=\begin{bmatrix} \frac{1}{3} & -\frac{2}{3} & -\frac{2}{3} \\ -\frac{2}{3} & \frac{1}{3} & -\frac{2}{3} \\ -\frac{2}{3} & -\frac{2}{3} & \frac{1}{3} \end{bmatrix} \tag{10.27}$$

For an ideal circulator the $\bar{S}$ matrix is

$$\bar{S}=\begin{bmatrix} 0 & S_{12} & 0 \\ 0 & 0 & S_{12} \\ S_{12} & 0 & 0 \end{bmatrix} \tag{10.28}$$

where S_{12} has unit amplitude.

The characteristic equation for this matrix is

$$-s_n^3+S_{12}^3=0 \tag{10.29}$$

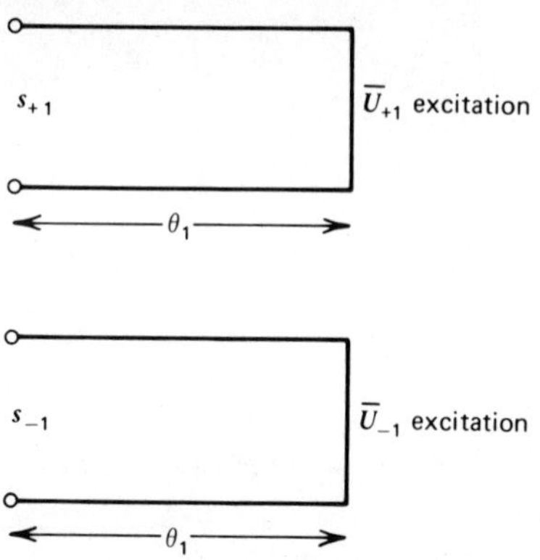

Figure 10.6. Eigennetworks of reciprocal 3-port junction.

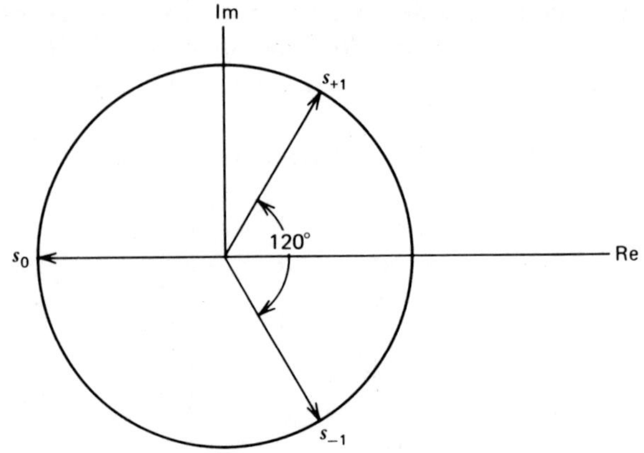

Figure 10.7. Eigenvalue diagram of ideal 3-port circulator.

For a circulator with $S_{12} = -1$ the result is

$$s_{+1} = e^{-j2(\theta_1 + \theta_{+1} + \pi/2)} \tag{10.30}$$

$$s_{-1} = e^{-j2(\theta_1 + \theta_{-1} + \pi/2)} \tag{10.31}$$

$$s_0 = e^{-j2\theta_0} \tag{10.32}$$

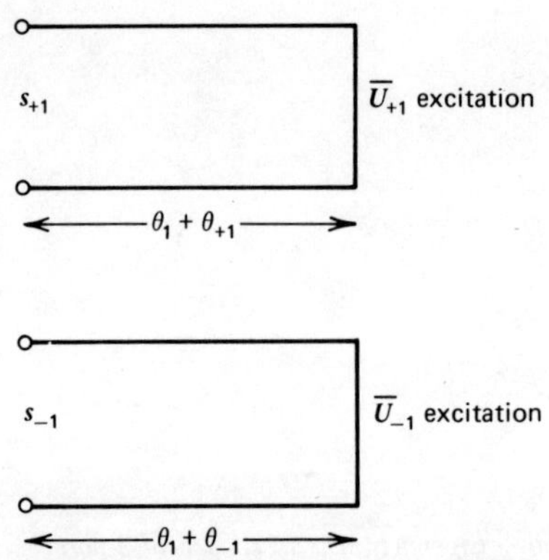

Figure 10.8. Eigennetworks of ideal 3-port circulator.

provided

$$\theta_1 = \theta_0 = \frac{\pi}{2} \tag{10.33}$$

$$\theta_{+1} = -\theta_{-1} = -\frac{\pi}{6} \tag{10.34}$$

The three reflection coefficients of the ideal circulator, therefore, lie equally spaced on the unit circle in the manner depicted in Figure 10.7. One set of 1-port eigennetworks that satisfy these reflection coefficients is shown in Figure 10.8.

In what follows it will be shown that the eigenvalues $s_{\pm}$ may be split in a magnetized ferrite since the corresponding eigenvectors $\overline{U}_{\pm}$ establish counterrotating magnetic fields at the center of the junction for which the scalar permeabilities $\mu_{\pm}$ of a magnetized ferrite apply.

10.3 *SCATTERING MATRIX EIGENVECTORS*

The eigenvectors of the junction are determined by the symmetry properties of the junction only. For a 3-port symmetric junction they are given by

$$\overline{U}_0 = \frac{1}{\sqrt{3}} \begin{bmatrix} 1 \\ 1 \\ 1 \end{bmatrix} \tag{10.35}$$

$$\overline{U}_{+1} = \frac{1}{\sqrt{3}} \begin{bmatrix} 1 \\ e^{-j2\pi/3} \\ e^{j2\pi/3} \end{bmatrix} \tag{10.36}$$

$$\overline{U}_{-1} = \frac{1}{\sqrt{3}} \begin{bmatrix} 1 \\ e^{j2\pi/3} \\ e^{-j2\pi/3} \end{bmatrix} \tag{10.37}$$

The above three eigenvectors represent the three possible ways of exciting the junction that give identical reflection coefficients at each port. Each excitation corresponds to one of the reflection eigenvalues s_0, s_{+1} and s_{-1} of the scattering matrix $\overline{S}$. This may be shown by solving the eigenvalue equations one at a time. Schematic diagrams for these three individual excitations are illustrated in Figures 10.9*a* through 10.9*c*. A linear combination of these excitations is equivalent to a single input at any port.

The nature of the electromagnetic fields at the center of the junction may be inferred by considering each eigenvector excitation one at a time.

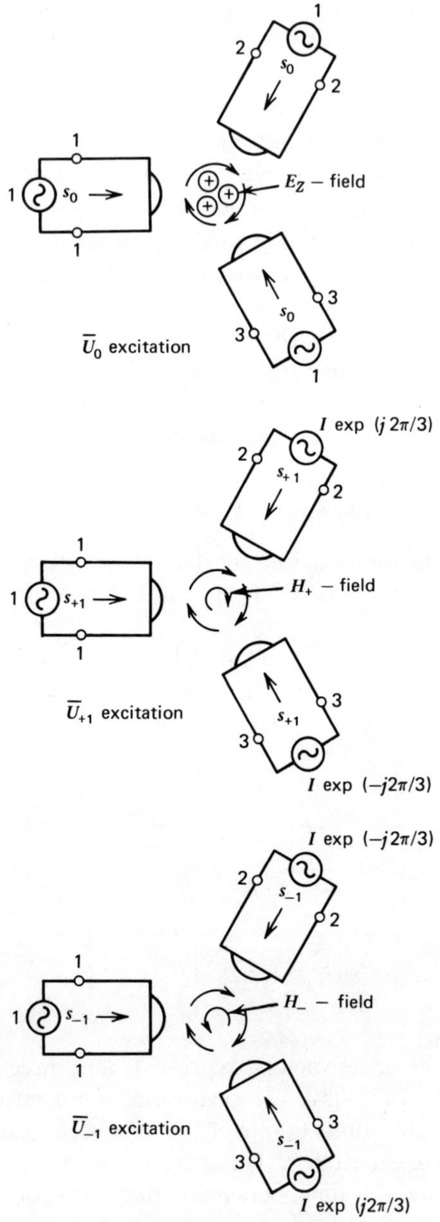

Figure 10.9. Schematic diagrams of eigenvector excitations showing electromagnetic fields at center of junction.

For the excitation corresponding to the eigenvector $\bar{U}_0$, the electromagnetic field at the center of the junction has only components parallel to the axis of the junction. For the eigensolutions corresponding to the eigenvectors $\bar{U}_{+1}$ and $\bar{U}_{-1}$, the axial fields vanish at the center of the junction, but the transverse components of the electric and magnetic field give rise to circularly polarized waves rotating in one sense for one eigenvector and in the opposite sense for the other. Corresponding to these three excitations, one obtains the reflection coefficients s_0, s_{+1}, and s_{-1} at any port.

The fields at the center of the junction in the transverse plane are given by straightforward vector addition by

$$\bar{h}_t = \bar{a}_x\left[\frac{\sqrt{3}}{2}(h_2 - h_3)\right] + \bar{a}_y\left[h_1 - \tfrac{1}{2}(h_2 + h_3)\right] \tag{10.38}$$

$$\bar{e}_z = \bar{a}_z[e_1 + e_2 + e_3] \tag{10.39}$$

For the $\bar{U}_0$ excitation

$$\bar{h}_t = 0 \tag{10.40}$$

$$\bar{e}_z = \sqrt{3}\,\bar{Q}_z \tag{10.41}$$

Thus the magnetic field for this excitation is zero at the center of the junction while the electric field is a maximum there. The equivalent 1-port circuit is therefore the opencircuited one used earlier to describe this excitation.

For the $\bar{U}_{\pm}$ excitations

$$\bar{h}_{t\pm} = \bar{a}_x\left(\pm j\frac{\sqrt{3}}{2}\right) + \bar{a}_y\left(\frac{\sqrt{3}}{2}\right) \tag{10.42}$$

$$\bar{e}_z = 0 \tag{10.43}$$

The two magnetic fields at the center of junction are counterrotating circularly polarized waves, whereas the electric fields are zero there. The equivalent shortcircuited transmission lines introduced earlier therefore apply.

Since the field patterns in the transverse plane for the $\bar{U}_{\pm}$ excitations are circularly polarized at the center of the junction they acquire the appropriate scalar permeabilities $\mu \mp K$ of the magnetized ferrite medium. Such a magnetized junction will therefore split the degeneracy between the reciprocal $s_{\pm}$ reflection coefficients of the junction. This arrangement can therefore be used to construct the scattering matrix of the ideal circulator established earlier.

10.4 *DIAGONALIZATION OF SCATTERING MATRIX*

If the eigenvalues are known it is possible to form the coefficients of the matrix $\bar{S}$. The relation between the two is obtained by diagonalizing $\bar{S}$, in the manner described in Chapter 1.

In this instance the matrix $\bar{U}$ is given in terms of the eigenvectors defined by Eqs. 10.35 through 10.37.

$$\bar{U}=\frac{1}{\sqrt{3}}\begin{bmatrix}1 & 1 & 1\\ 1 & e^{-j2\pi/3} & e^{j2\pi/3}\\ 1 & e^{-j4\pi/3} & e^{j4\pi/3}\end{bmatrix} \tag{10.44}$$

The matrix $(\bar{U}^*)^T$ is

$$(\bar{U}^*)^T=\frac{1}{\sqrt{3}}\begin{bmatrix}1 & 1 & 1\\ 1 & e^{j2\pi/3} & e^{j4\pi/3}\\ 1 & e^{-j2\pi/3} & e^{-j4\pi/3}\end{bmatrix} \tag{10.45}$$

In the case of the 3-port junction, there are a degenerate pair of eigenvalues and one nondegenerate one. The diagonal matrix $\bar{\lambda}$ with the eigenvalues of $\bar{S}$ is, therefore,

$$\bar{\lambda}=\begin{bmatrix}s_0 & 0 & 0\\ 0 & s_{+1} & 0\\ 0 & 0 & s_{-1}\end{bmatrix} \tag{10.46}$$

Diagonalizing the $\bar{S}$ matrix gives the following relation between the scattering coefficients and the eigenvalues

$$3S_{11}=s_0+s_{+1}+s_{-1} \tag{10.47}$$

$$3S_{12}=s_0+s_{+1}e^{j2\pi/3}+s_{-1}e^{-j2\pi/3} \tag{10.48}$$

$$3S_{13}=s_0+s_{+1}e^{-j2\pi/3}+s_{-1}e^{j2\pi/3} \tag{10.49}$$

The eigenvalues given by Eqs. 10.30 through 10.32 give the conditions for an ideal circulator. If s_{+1} and s_{-1} are interchanged, the sense of circulation is reversed. The eigenvalues become interchanged when the biasing direct magnetic field is reversed.

It can be seen from Eq. 10.47 that the spur of the scattering matrix is equal to the sum of the eigenvalues. This is a general result.

10.5 *LUMPED ELEMENT CIRCULATOR*

One circulator geometry which can be readily analyzed in terms of the eigenvalues of the ideal circulator is the lumped-element configuration depicted in Fig. 10.10. It consists of a ferrite disk with three coils wound on it so that the r.f. magnetic fields of the coils are oriented at 120 deg with respect to each other. A d.c. magnetic field is applied normal to the plane of the disk. This symmetrical but nonreciprocal element can be used to form a circulator by connecting capacitors either in series or shunt with the load and source impedances.

One configuration of the shunt arrangement consists of a ferrite disk at the junction of a mesh arrangement of three short sections of short-circuited stripline at 120 deg that are insulated from each other. This geometry is shown in Fig. 10.11. If the shortcircuited striplines are electrically short the energy within the disk geometry is essentially magnetic. The junction degenerate resonances are then obtained by connecting shunt capacitances at the three terminals, whereas the impedance level of the device is obtained by adjusting the d.c. magnetic field. The voltage current relations at the terminals of the mesh arrangement are given by

$$\begin{bmatrix} V_1 \\ V_2 \\ V_3 \end{bmatrix} = j\omega L_0 \begin{bmatrix} \mu_i & \frac{-\mu_i}{2} & \frac{-\mu_i}{2} \\ \frac{-\mu_i}{2} & \mu_i & \frac{-\mu_i}{2} \\ \frac{-\mu_i}{2} & \frac{-\mu_i}{2} & \mu_i \end{bmatrix} \begin{bmatrix} I_1 \\ I_2 \\ I_3 \end{bmatrix} \tag{10.50}$$

Here L_0 is the inductance of the constituent mesh, and μ_i is the initial permeability of the ferrite disk.

The impedance eigenvalues for this network are

$$z_0 = 0 \tag{10.51}$$

$$z_{+1} = \frac{j3\omega L_0 \mu_i}{2} \tag{10.52}$$

$$z_{-1} = \frac{j3\omega L_0 \mu_i}{2} \tag{10.53}$$

The impedance eigenvalues consist of one nondegenerate eigenvalue and a pair of degenerate ones.

The admittance eigenvalues for this network are the reciprocal of the

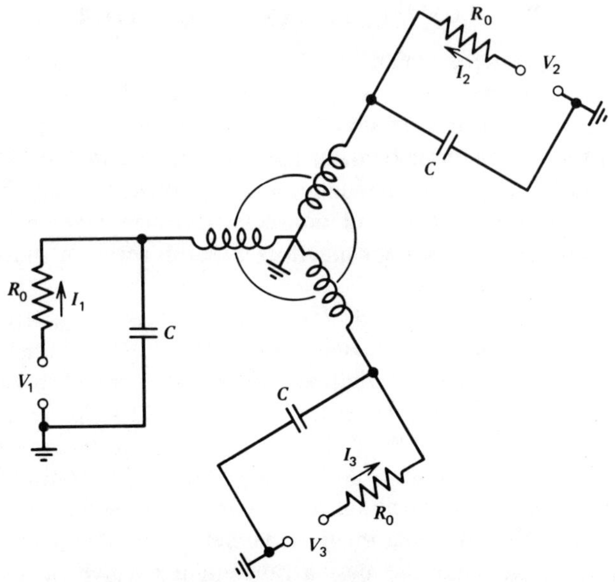

Figure 10.10. Schematic of LC-lumped element circulator.

impedance ones

$$y_0 = \infty \tag{10.54}$$

$$y_{+1} = \frac{2}{j3\omega L_0 \mu_i} \tag{10.55}$$

$$y_{-1} = \frac{2}{j3\omega L_0 \mu_i} \tag{10.56}$$

The above admittance eigenvalues can be made to coincide with those of the first circulation adjustment by adding shunt capacitances at each terminal.

$$y_0 = \infty \tag{10.56}$$

$$y_{+1} = j\omega C + \frac{2}{j3\omega L_0 \mu_i} = 0 \tag{10.57}$$

$$y_{-1} = j\omega C + \frac{2}{j3\omega L_0 \mu_i} = 0 \tag{10.58}$$

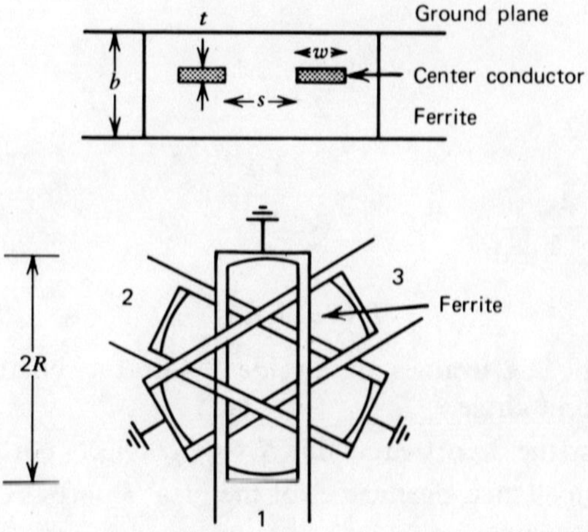

Figure 10.11. Geometry of LC-lumped element circulator using shortcircuited striplines.

This adjustment establishes the isotropic junction resonances. The second circulation adjustment is now obtained by splitting the degenerate admittance eigenvalues until they coincide with those of an ideal circulator by making use of the scalar permeabilities defined in Chapter 8.

$$y_0 = \infty \tag{10.59}$$

$$y_{+1} = j\omega C + \frac{2}{j3\omega L_0(\mu - K)} = -j\frac{Y_0}{\sqrt{3}} \tag{10.60}$$

$$y_{-1} = j\omega C + \frac{2}{j3\omega L_0(\mu + K)} = j\frac{Y_0}{\sqrt{3}} \tag{10.61}$$

The boundary conditions on the admittance eigenvalues follow directly from those of the scattering eigenvalues in Eqs. 10.30 through 10.34. The two circulation conditions are therefore

$$Y_0 = \frac{\sqrt{3}}{\omega L} \cdot \frac{K}{\mu} \tag{10.62}$$

and

$$\omega^2 LC = 1 \tag{10.63}$$

where

$$L = \frac{3}{2}\,\mu_e L_0 \tag{10.64}$$

$$\mu_e = \frac{\mu^2 - K^2}{\mu} \tag{10.65}$$

PROBLEMS

1. Show that the eigenvalues of an ideal 4-port circulator lie equally spaced on a unit circle.
2. Verify that for the 3-port circulator $S_{12} = 1$ connects ports 2 to 1.
3. Obtain the impedance eigenvalues of the ideal 3-port circulator from its scattering eigenvalues.
4. Derive the impedance matrix of an ideal circulator in terms of its eigenvalues.
5. Using the unitary condition show that $|S_{11}| \approx |S_{13}|$ for $S_{12} \approx 1$.
6. Show that Eqs. 10.35 through 10.37 satisfy Eq. 10.14.
7. Verify that Eqs. 10.60 through 10.62 are consistent with Eqs. 10.30 through 10.34.
8. Construct the scattering matrix of the lumped element circulation.

FURTHER READING

1. H. Fowler, presented at the 1950 Symposium on Microwave Properties and Applications of Ferrites, Harvard University, Cambridge, Mass.
2. H. N. Chait and T. R. Curry, *Y*—Circulator, *J. Appl. Phys.*, **30** (1959), p. 152.
3. F. M. Aitken and R. McLean, Some properties of the waveguide *Y* circulator, *Proc. IEEE*, **110** (2) (1963), pp. 256–260.
4. B. A. Auld, The synthesis of symmetrical waveguide circulators, *IRE Trans. Microwave Theory Tech.*, **7** (1959), pp. 238–246.
5. C. E. Fay and R. L. Comstock, Operation of the ferrite junction circulator, *IEEE Trans. Microwave Theory Tech.*, **13**, (1965), p. 15.
6. J. Helszajn, *Nonreciprocal microwave junctions and circulators*, Wiley, New York, 1976.

CHAPTER ELEVEN

Variable Capacitance Diode Circuits

The varactor or variable capacitance diode is a semiconductor device characterized by a voltage variable capacitance. This diode is widely used to tune the resonant circuit of bulk semiconductor oscillators, to construct semiconductor phase shifters, and finds application as a nonlinear capacitance in the construction of frequency multipliers and parametric amplifiers. It consists of a $P+NN+$ or $N+PP+$ layer sequence as illustrated in Figure 11.1. In such an unbiased semiconductor diode, the mobile charge carriers move away from the immediate vicinity of the *pn* junction, leaving a narrow region of uncompensated fixed charges called the depletion or space charge layer. If the diode is biased in the reverse direction, the mobile charge carriers move further away from the junction and so widen the depletion layer. This effect gives rise to a junction capacitance that is voltage dependent and the diode, therefore, can be used as a voltage variable capacitor. As indicated in Figure 11.2, the device can be considered as a parallel-plate capacitor, the depletion layer being the dielectric.

Figure 11.3 depicts a simplified equivalent circuit for a packaged varactor diode. L_p and C_p are determined by the package design that is similar to that of the *p-i-n* diode in Chapter 12. The series resistance R_s and the voltage-dependent capacitance $C_j(V)$ are both functions of the semiconductor material. Typical varactor packages are shown in Figure 11.4.

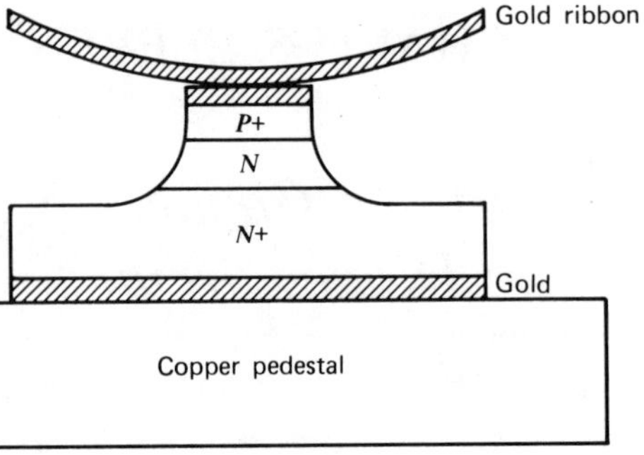

Figure 11.1. Construction of a varactor diode.

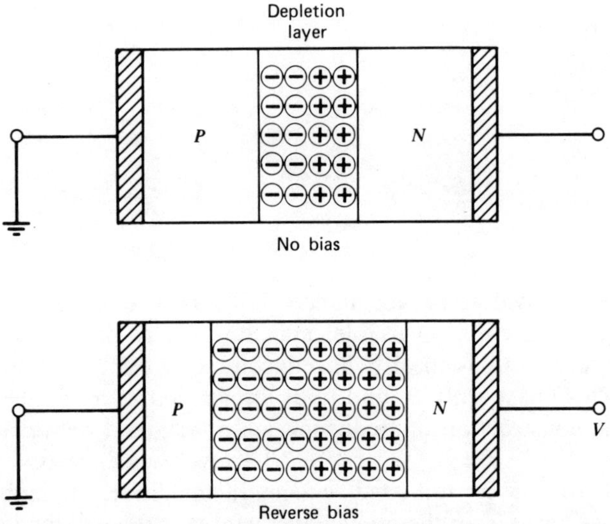

Figure 11.2. Depletion layer model showing charge distribution of

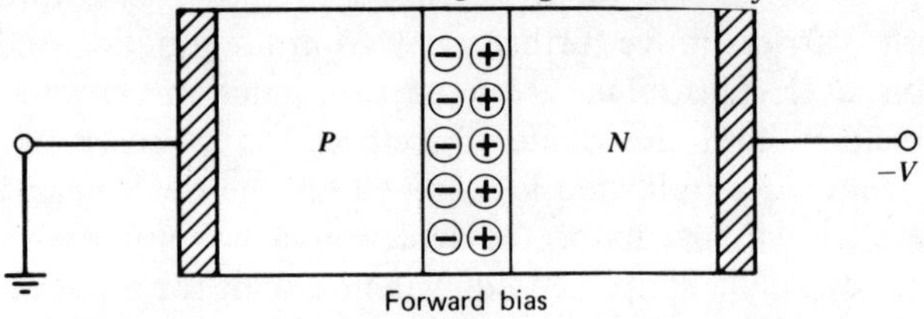

Figure 11.2. Depletion layer model showing charge distribution of ionized donors and acceptors under forward, reverse, and zero bias.

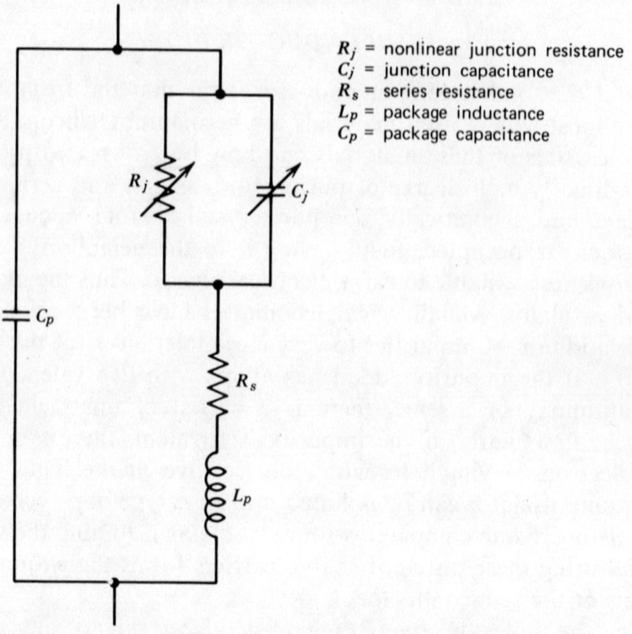

Figure 11.3. Simplified equivalent circuit of packaged varactor diode.

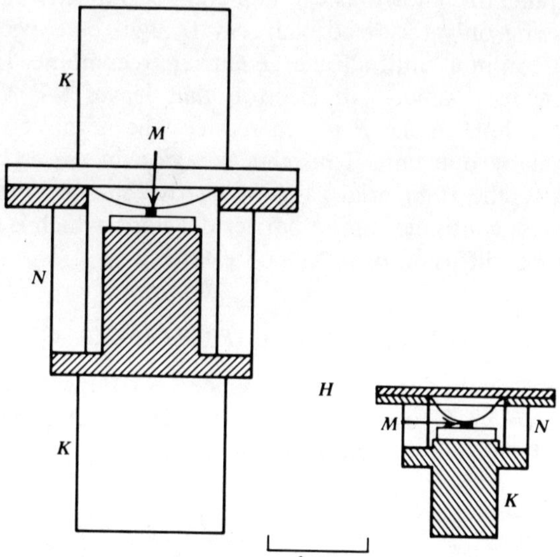

Figure 11.4. Varactor diode packages (courtesy Associated Semiconductor Manufacturers Ltd., Wembley, England). K, kovar; stud. M, diffused mesa; N, ceramic insulator; H, top cap.

11.1 *SEMICONDUCTOR DIODES*

The basis of the solid-state semiconductor is the material from which it is made. The most widely used materials are germanium, silicon, and GaAs and single crystals of these materials can now be grown and processed to an extraordinarily high degree of purity. Both silicon and germanium are quadrivalent, and, theoretically, in a pure crystal the four valency electrons of each atom are occupied, tightly tying it to the neighboring atom. No free electrons are available to carry electrical charge. Thus the material is a very good insulator. Metallurgical techniques have been developed that permit the addition of impurities to very close tolerances (of the order of 1 part in 10^8). If the impurity added has atoms with five valency electrons such as antimony or arsenic, there is a surfeit of unattached negative electrons, and, similarly, if the impurity is trivalent, there is a deficit of negative electrons leaving holes with a net positive charge. Thus the silicon or germanium insulator can be polluted into an *n*-type or *p*-type conductor (albeit a mediocre one compared with most metals). Joining the two pieces of material using these different charge carriers forms the *pn*-junction that is the basis of the semiconductor diode.

Initially, the holes in the *P* region tend to spread out or diffuse throughout the crystal, a process that can be expected whenever the density of carriers is larger at one plane than at another. Likewise, electrons from the *N* region diffuse throughout the crystal. In the narrow region of the junction, known as the depletion region, the mobile charges recombine leaving only the fixed charges. It might be expected that this process would continue until all charge carriers recombine. However, this is not the case in practice. An electron that leaves the *N* region and combines with a hole in the *P* region reduces the negative charge in the *N*-type material by one unit. Thus the *N* region increases by a positive charge. Similarly, the *P* material gains a negative charge from its loss of a hole. This process continues until a barrier develops, which is large enough to prevent further diffusion of holes and electrons.

11.2 *JUNCTION CAPACITANCE*

The dependence of the junction capacitance $C_j(V)$ upon the reverse bias voltage is shown in Figure 11.5 for a typical varactor diode. A good representation for this diode characteristic is

$$C_j(V) = \frac{C_j(0)}{[1+(V/\phi)]^{\gamma}} \tag{11.1}$$

$C_j(0)$ is the junction capacitance at zero bias, V is the applied reverse

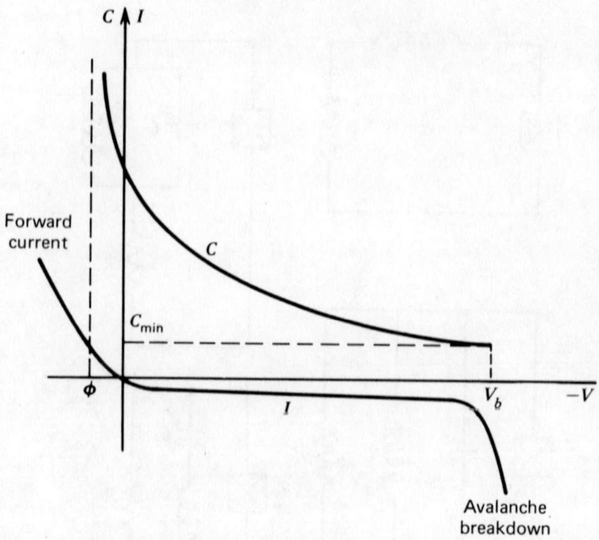

Figure 11.5. Typical current and capacitance relationships for varactor diode.

voltage, and ϕ is the contact potential that is a function of the type of semiconductor and the level of doping. The exponent γ is a function of the impurity profile of the junction and has a value of $\frac{1}{2}$ for abrupt junctions, and $\frac{1}{3}$ for linear graded ones. $C_j(V)$ is also sometimes written as

$$C_j(V) = C_{\min}\left(\frac{\phi + V_b}{\phi + V}\right)^{\gamma} \tag{11.2}$$

where $C_{\min}$ is the capacitance at the reverse breakdown voltage V_b

$$C_{\min} = \frac{C_j(0)\phi^{\gamma}}{(\phi + V_b)^{\gamma}} \tag{11.3}$$

A figure of merit called the cut-off frequency can be calculated in terms of R_s and $C_j(V)$

$$\omega_c = \frac{1}{R_s C_j(V)} \tag{11.4}$$

This defines the frequency at which the Q factor of the diode is equal to unity. The Q factor at the operating frequency is then defined as

$$\frac{\omega_c}{\omega_o} = Q_o \tag{11.5}$$

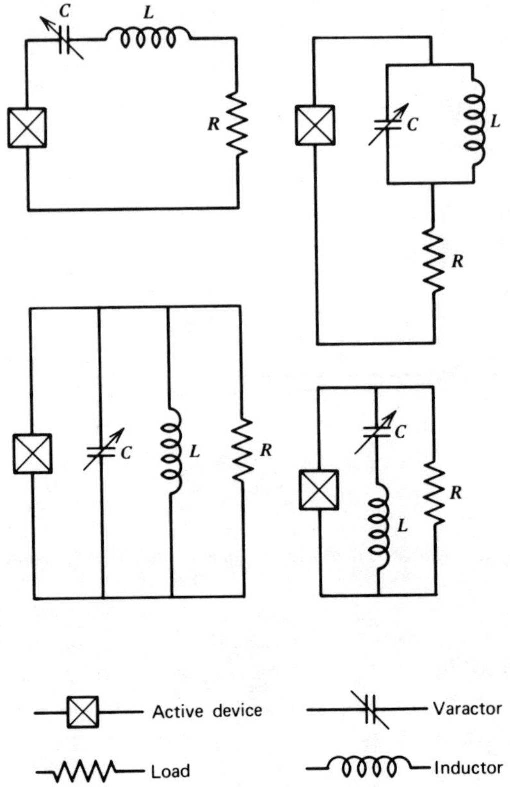

Figure 11.6. Some wide-band varactor circuits used to tune bulk devices.

Figure 11.6 indicates a number of ways a varactor diode can be incorporated into an oscillator to tune a bulk device such as a Gunn diode.

11.3 *CAPACITANCE, ELASTANCE, AND CHARGE*

The voltage-dependent nonlinear capacitance defined by Eq. 11.1 may be expanded into a periodic function of ω when driven by a single signal at frequency ω. To show this it is only necessary to replace V by $V_0 \sin \omega t$ in Eq. 11.2. The result is

$$C_j(V) = C_{\min}\left(\frac{\phi + V_b}{\phi + V_0 \sin \omega t}\right)^{\gamma} \tag{11.6}$$

Rearranging this equation yields

$$C_j(V) = C_{\min}\left(\frac{\phi + V_b}{\phi}\right)^{\gamma} \frac{1}{\left[1 + \frac{V_0}{\phi}\sin\omega t\right]^{\gamma}} \tag{11.7}$$

Expanding this function as a binomial series gives

$$C_j(V) = C_{\min}\left(\frac{\phi + V_b}{\phi}\right)^{\gamma}\left[1 - \gamma\left(\frac{V_0}{\phi}\sin\omega t\right) - \gamma\frac{(\gamma-1)}{2!}\left(\frac{V_0}{\phi}\sin\omega t\right)^2\right] \tag{11.8}$$

Thus the general form for $C_j(V)$ is

$$C_j(V) = C_0 + C_1\sin\omega t + C_2\sin 2\omega t + C_3\sin 3\omega t + \ldots \tag{11.9}$$

It is sometimes useful to define an elastance $S = 1/C$. Using 11.6 gives

$$S(V) = S_{\max}\left(\frac{\phi + V_0\sin\omega t}{\phi + V_b}\right)^{\gamma} \tag{11.10}$$

Writing this function as a binomial series leads to

$$S(V) = S_0 + S_1\sin\omega t + S_2\sin 2\omega t + S_3\sin 3\omega t + \ldots \tag{11.11}$$

It is readily shown that the charge across the diode is also a periodic function of ω. The relation between charge, voltage and capacitance is

$$Q(V) = \int C(V)\,dV \tag{11.12}$$

Integrating Eq. 11.2

$$Q(V) = C_{\min}\frac{(\phi + V_b)^{\gamma}}{(1-\gamma)}(\phi + V)^{-(\gamma-1)} \tag{11.13}$$

Replacing V by $V_0\sin\omega t$ and expanding the above function into a binomial series immediately gives the desired result

$$Q(V) = Q_0 + Q_1\sin\omega t + Q_2\sin 2\omega t + Q_3\sin 3\omega t + \ldots \tag{11.14}$$

11.4 *FREQUENCY MULTIPLIERS*

The periodic nature of the charge across the diode immediately leads to the principle of frequency multiplication. The current through the diode is given in terms of the charge by

$$i = \frac{dQ(V)}{dt} \tag{11.15}$$

Using the periodic form for $Q(V)$ given by Eq. 11.14 gives

$$i = \omega Q_1 \cos \omega t + 2\omega Q_2 \cos 2\omega t + \ldots. \tag{11.16}$$

Thus the diode current consists of a fundamental frequency component plus higher order harmonic terms.

Figure 11.7 shows a simple type of doubler circuit. This is merely a combination of two resonant circuits coupled together through a common impedance that is the diode itself. Ordinarily the varactor impedance is small compared to the reactances at undesired frequencies in either of the two separate loops, so there is a natural division of current. The major portion of the fundamental current flows in the first loop, and substantially all of the second harmonic current flows in the second loop. Coupling links are added to the two inductances for input and output connections.

Triplers and quadrupler circuits are somewhat similar to the doubler circuit configuration. Figure 11.8 shows one such arrangement. It consists of three resonant circuits again coupled through a common varactor. One loop, resonant with the diode at the fundamental, is coupled to the input. The second loop, resonant with the diode at the second harmonic frequency is added in shunt with the varactor. This is the idler frequency, which has been shown to be necessary to achieve maximum efficiency by

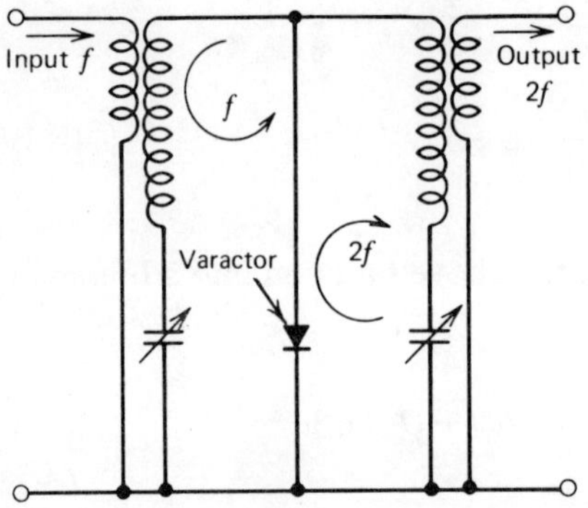

Figure 11.7. Varactor doubler circuit.

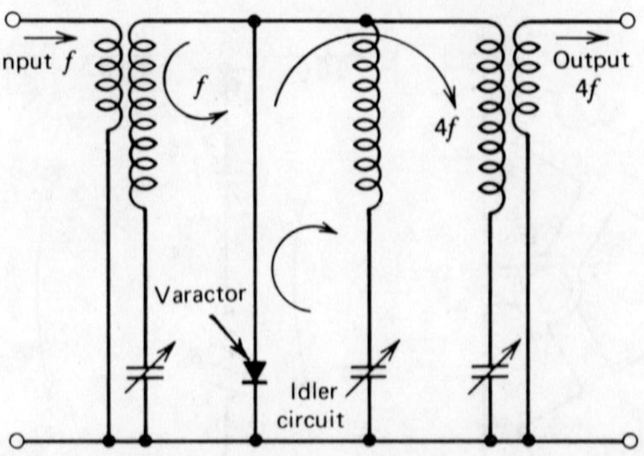

Figure 11.8. Varactor quadrupler circuit.

allowing the second harmonic current to flow. The third loop is resonant with the diode at either the third or fourth harmonic and is coupled to the output.

11.5 *THE PARAMETRIC AMPLIFIER*

The parametric amplifier to be analyzed is the 1-port negative resistance type in which the input and output signals are separated by a circulator. A practical amplifier suitable for analysis is shown in Figure 11.9 (the pump circuit is omitted in this schematic). Parametric amplifiers have two input frequencies ω_s and ω_p, and the varactor is used to convert energy at the pump frequency ω_p to provide low noise amplification at the signal frequency ω_s. In addition to these two frequencies one other resonant circuit at ω_i must be established. This third tank circuit is known as an idler one. The relation between the three frequencies and power levels in a nonlinear reactance have been given by Manley and Rowe.

$$\frac{P_s}{\omega_s} = \frac{P_i}{\omega_i} = -\frac{P_p}{\omega_p} \tag{11.17}$$

$$\omega_p = \omega_s + \omega_i \tag{11.18}$$

The starting point for the analysis of the parametric amplifier is to write the voltage across the varactor for the circuit in Figure 11.9.

$$V(t) = \int S(t)i(t)dt + R_s i(t) \tag{11.19}$$

The small signal assumption made here is that the elastance time function is strictly controlled by the pumping and that the small signal currents do

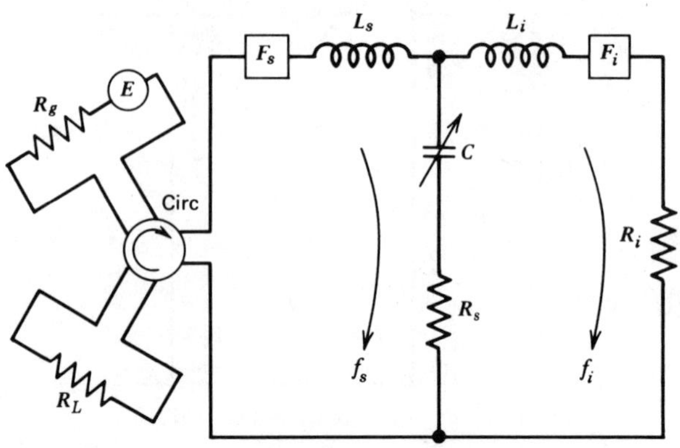

Figure 11.9. Schematic of parametric amplifier.

not effect it. Thus $S(t)$ may be taken outside the integral sign

$$V(t) = S(t)\int i(t)dt + R_s i(t) \tag{11.20}$$

$V(t)$, $i(t)$, and $S(t)$ are now all expanded in Fourier series and the resonant terms at ω_s, ω_p, and ω_i are retained

$$i_s(t) = i_s \exp(j\omega_s t) + i_s^* \exp(-j\omega_s t) \tag{11.21}$$

$$i_i(t) = i_i \exp(j\omega_i t) + i_i^* \exp(-j\omega_i t) \tag{11.22}$$

$$S(t) = S_0 + S_1 \exp(j\omega_p t) + S_1^* \exp(-j\omega_p t) \tag{11.23}$$

Substituting the above relations into the varactor equation and equating synchronous terms gives

$$\begin{bmatrix} V_s \\ V_i^* \end{bmatrix} = \begin{bmatrix} \dfrac{S_0}{j\omega_s} + R_s & -\dfrac{S_1}{j\omega_i} \\ \dfrac{S_1}{j\omega_s} & -\dfrac{S_0}{j\omega_i} + R_s \end{bmatrix} \begin{bmatrix} i_s \\ i_i^* \end{bmatrix} \tag{11.24}$$

This equation indicates that the signal current flowing through the time-varying elastance gives rise to the idler voltage, whereas the idler current through the time-varying elastance produces a signal voltage component.

The 1-port device can operate as an amplifier if and only if the signal impedance has a negative real part. The input impedance of the amplifier with series tuned signal and idler circuits is obtained by eliminating the idler quantities. Making use of the idler circuit equation at resonance gives

$$V_i = R_i i_i \tag{11.25}$$

where R_i is the external circuit resistance.

Substituting this last equation into Eq. 11.24 and setting the imaginary parts of the signal input impedance to zero gives

$$R_- \approx R_s\left[1 - \frac{S_1^2}{\omega_s \omega_i R_s (R_s + R_i)}\right] \tag{11.26}$$

Thus amplification will occur provided

$$1 \leqslant \frac{S_1^2}{\omega_s \omega_i R_s (R_s + R_i)} \tag{11.27}$$

The above equation is also sometimes written in terms of a modulation index m_1 and a varactor cut-off frequency ω_c

$$m_1 = \frac{S_1}{S_{\max} - S_{\min}} \tag{11.28}$$

$$\omega_c = \frac{S_{\max} - S_{\min}}{R_s} \tag{11.29}$$

Equation 11.29 reduces to Eq. 11.4 when $S_{\min} = 0$.

By extending the definition of the reflection coefficient to include negative terminating resistances, the gain of the negative-resistance amplifier and circulator depicted in Figure 11.9 can be written as

$$G = \left(\frac{|R_-| + R_0}{-|R_-| + R_0}\right)^2 \tag{11.30}$$

where $|R_-|$ is the magnitude of the negative resistance.

11.6 *VARACTOR ANALOGUE PHASE SHIFTER*

Microwave control circuits are fully discussed in Chapter 12 that deals with *p-i-n* variable-resistance diode devices. However, the variable-capaci-

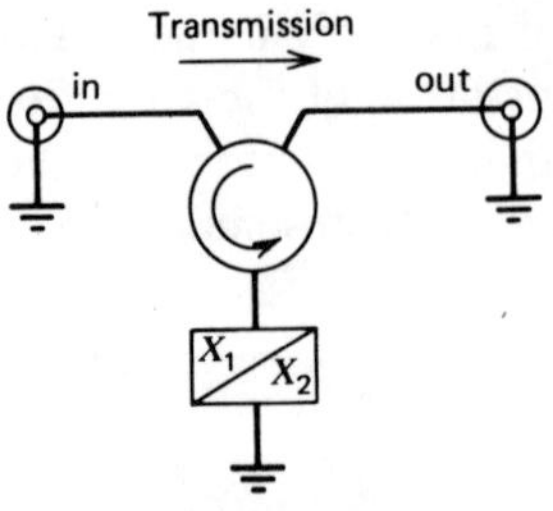

Figure 11.10a. Varactor diode using circulator.

tance diode may also be used in conjunction with circulators and hybrid networks for the construction of reflection-type analog phase shifters. Two such phase shifters are illustrated in Figure 11.10*a* and 11.10*b*.

The reflection coefficient for a varactor diode terminating a transmission line of characteristic impedance Z_0 is

$$S_{\text{in}} = \frac{Z_{\text{in}} - Z_0}{Z_{\text{in}} + Z_0} \tag{11.31}$$

If the package parasitics are neglected

$$Z_{\text{in}} = \frac{1}{j\omega C(V)} \tag{11.32}$$

Thus

$$S_{\text{in}} = -\exp(-j2\theta) \tag{11.33}$$

where

$$\tan 2\theta = \omega C(V) Z_0 \tag{11.34}$$

The reactance curve obtained in this way can be adjusted to closely fit a

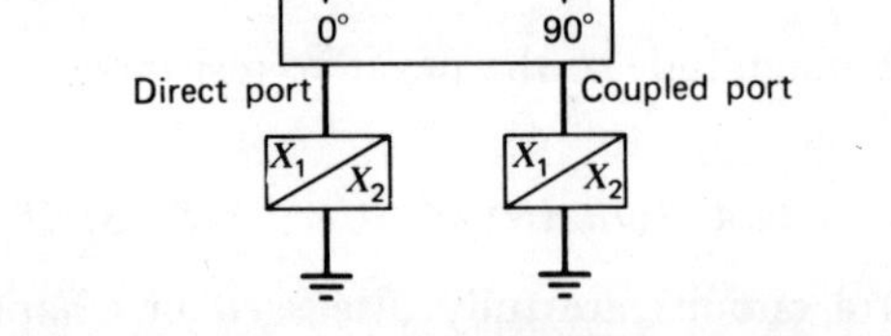

Figure 11.10b. Varactor diode phase shifter using hybrid circuit.

tangent curve over 90 deg by adding a series inductance to the varactor diode. This condition is obtained by satisfying the following relation

$$Z_0 \tan\left(\frac{2\pi l}{\lambda}\right) = \omega L - \frac{1}{\omega C(V)} \tag{11.35}$$

The input/output relation of the hybrid arrangement in Figure 11.10*b* for a single input at port 1 is given by

$$\begin{vmatrix} b_1 \\ b_2 \\ b_3 \\ b_4 \end{vmatrix} = \frac{1}{\sqrt{2}} \begin{vmatrix} 0 & 0 & 1 & j \\ 0 & 0 & j & 1 \\ 1 & j & 0 & 0 \\ j & 1 & 0 & 0 \end{vmatrix} \begin{vmatrix} 1 \\ 0 \\ -b_3 \\ -b_4 \end{vmatrix} \tag{11.36}$$

This relation is fully developed for the *p-i-n* variable-resistance diode in Chapter 12.

A 2-port transmission phase network can also be constructed by periodically loading a transmission line with varactor diodes, but this configuration requires a large number of diodes to obtain 180-deg phase modulation.

PROBLEMS

1. An abrupt diode with $\gamma = \frac{1}{2}$ has a junction capacitance of $C_j(0) = 0.25$ $\mu\mu$F. Find $C_j(V_b)$ at the reverse breakdown voltage $V_b = -6$ V for $\phi = 8$ V.
2. Find the coefficients C_0, C_1, and C_2 for $V_0 = 1$ V for the diode in example 1. Repeat for $V_0 = 2$ V.
3. Find the elastances S_0, S_1, and S_2 for the diode in problem 1, for $V_0 = 1$ V and 2 V.
4. Obtain the charge coefficients for the diode in problem 1 for V_0 given in problem 2.
5. A parametric amplifier has $\omega_S = 13$ GHz, $\omega_p = 35$ GHz, $R_s = 1$ Ω, $R_i = 1$ Ω, and S_1 in problem 2. Find its input impedance R_-.
6. Obtain the gain of a parametric amplifier having the input impedance in problem 5.
7. Construct the current waveform given by Eq. 11.16 using the charge coefficients in problem 4.
8. Determine the idler frequency for the parametric amplifier in problem 5.

9. An analogue phase shifter is switched between $C_j(0)$ and $C_j(V_b)$. Using $C_j(0)$ and $C_j(V_b)$ in problem 1, calculate the phase angle $\theta(0)-\theta(V_b)$.
10. An unpackaged diode with a series resistance $R_s = 1\ \Omega$ is used to terminate an ideal circulator. Calculate its reflection coefficients at 9 GHz for $C_j(0)$ and $C_j(V_b)$ in problem 1.

FURTHER READING

1. P. Penfield and R. P. Rafuse, *Varactor Applications*, MIT Press, Cambridge, Mass., 1962.
2. A. Uhlir, Jr., The potential of semiconductor diodes in high frequency communications, *Proc. IRE* (June 1958).
3. R. L. Camisa, B. F. Hitch, S. Yuam, and M. Ettenberg, Microwave integrated circuit M.I.S. varactor phase shifter, *RCA Review*, **36**, 2 (June 1975).
4. B. F. Nicholson, A *Q* band high level varactor upconverter, *Marconi Rev.* **38**, 196 (1975).
5. D. G. Vice, parametric amplifier noise, *IEEE Trans. Microwave Theory Tech.* (March 1965).

CHAPTER TWELVE

PIN Control Devices

The *p-i-n* (PIN) is a microwave semiconductor diode that can be used as an electrically variable resistor. It can be incorporated in waveguide, coaxial, and stripline components for microwave switching, attenuation, limiting, modulation, and phase shifting. The PIN diode consists of a high-resistivity semiconductor or *I* (intrinsic) layer between two highly doped P+ and N+ regions as shown in Figure 12.1. The variable element is a resistance R_i that is related to the *I* layer thickness. This layer has a

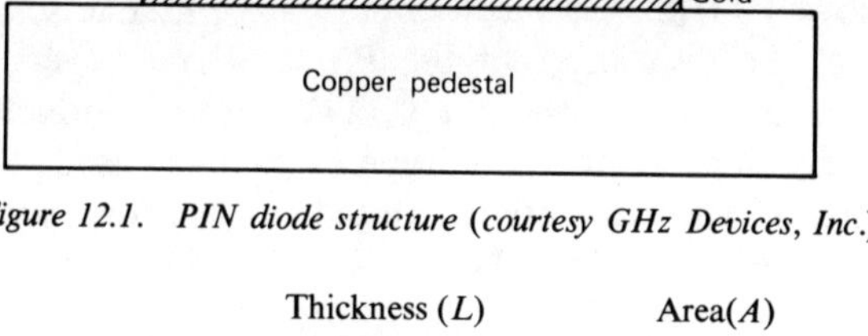

Figure 12.1. PIN diode structure (courtesy GHz Devices, Inc.).

	Thickness (L)	Area(A)
$P+$layer	0.76×10^{-3} cm	2×10^{-3} cm^2
I layer	7.6×10^{-3} cm	3.12×10^{-3} cm^2
$N+$layer	10.2×10^{-3} cm	4.5×10^{-3} cm^2
Metallization	0.127×10^{-3} cm	4.5×10^{-3} cm^2
Pedestal	10.2×10^{-2} cm	12.9×10^{-3} cm^2

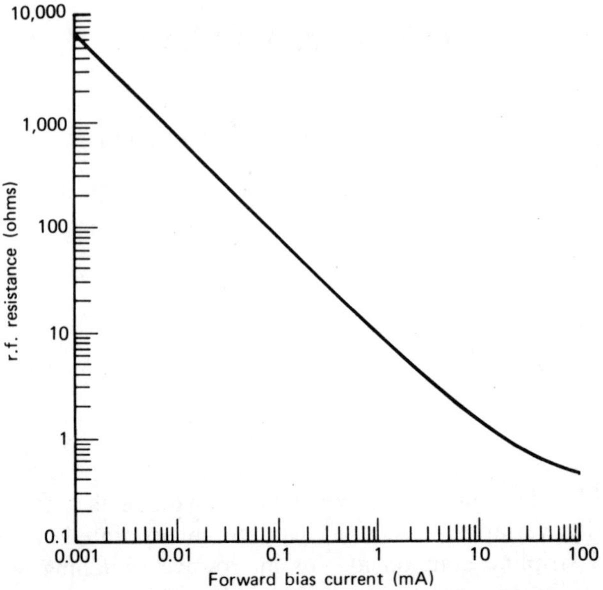

Figure 12.2. R.f. resistance versus forward current for PIN diode.

very large resistivity that can be increased still further by reverse biasing the diode but that decreases rapidly when the diode is forward biased as mobile carriers from the *P* and *N* regions are injected into the *I* layer. A typical resistance-versus-current curve is illustrated in Figure 12.2. Figure 12.3 depicts some typical microwave diode packages. A simplified circuit of a packaged diode is given in Figure 12.4. The resistance R_i is the variable element and has been mentioned already. The junction capacitance C_i is the capacitance of the *I* layer and is essentially constant at frequencies above 1 GHz. A typical value is 0.2 pF. The series resistance R_s is the sum of the resistances of the $P+$ and $N+$ layers and contact resistance, R_s has a value of about 1 Ω. The package inductance L_p and capacitance C_p are dependent on the type of package used. These may be omitted when the diode is directly mounted in chip form into the microwave circuit. Figure 12.5 shows the equivalent circuits which apply with the diode zero and forward biased.

PIN control circuits may be easily understood by establishing the input/output waves of the control circuits in terms of scattering matrices. This is the approach used in what follows to analyze some of the more important PIN control circuits.

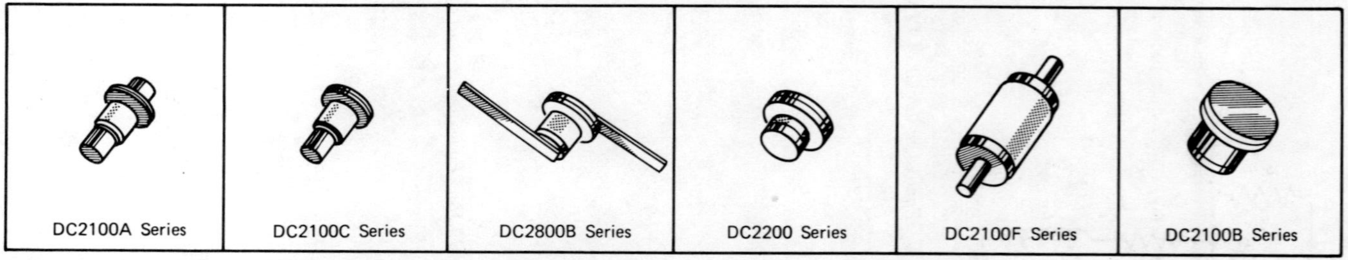

Figure 12.3. PIN diode packages (courtesy AEI Semiconductors, Ltd.).

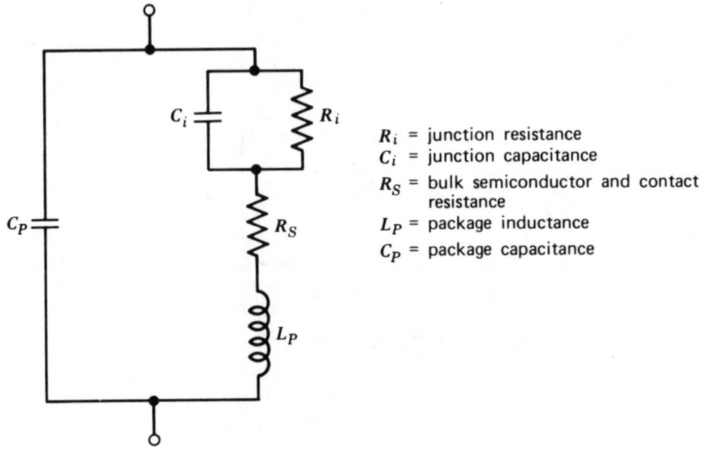

Figure 12.4. Equivalent circuit of PIN diode.

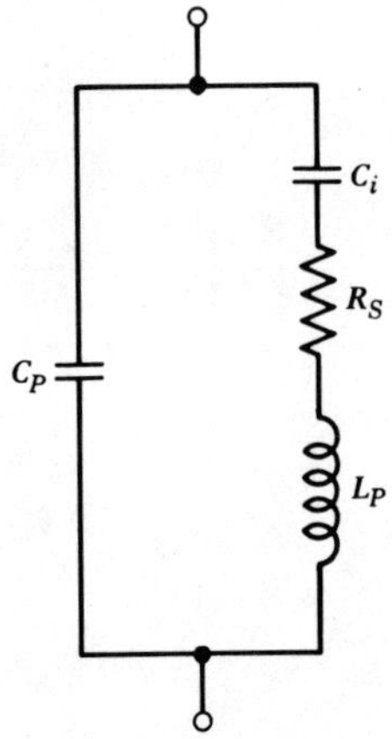

Figure 12.5a. Equivalent circuit of PIN diode for zero or reverse bias.

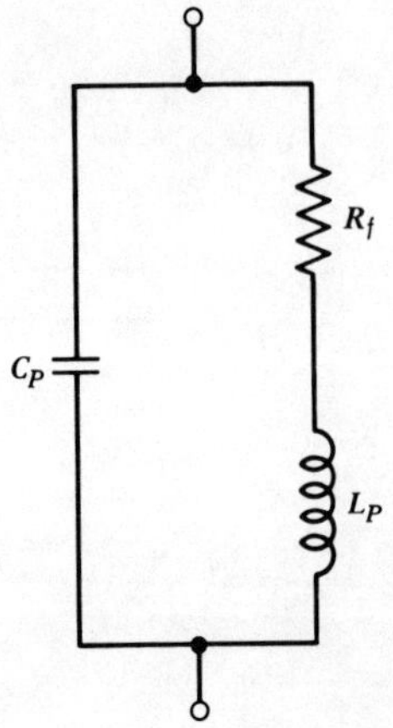

Figure 12.5b. Equivalent circuit of PIN diode for forward bias.

12.1 *BASIC SWITCHING CIRCUITS*

The two basic switch circuits are shown in Figures 12.6 and 12.7. The shunt diode switch is on when the diode is zero or reversed biased, whereas it is off when the diode is forward biased. The equivalent circuits that apply at the two bias voltages are illustrated in Figure 12.5. The operation of the series diode configuration is the dual of the shunt device.

The transmission and reflection losses of the respective switches are best expressed in terms of their scattering matrices.

$$\bar{S}=\frac{1}{y+2}\begin{bmatrix} -y & 2 \\ 2 & -y \end{bmatrix} \tag{12.1}$$

and

$$\bar{S}=\frac{1}{z+2}\begin{bmatrix} z & 2 \\ 2 & z \end{bmatrix} \tag{12.2}$$

where y and z are the normalized shunt and series immittances of the diode networks for the diodes zero or forward biased.

The insertion loss for the shunt device is

$$L(\text{dB})=10\log_{10}\frac{1}{|S_{21}|^2}=10\log_{10}\left|1+\frac{y}{2}\right|^2 \tag{12.3}$$

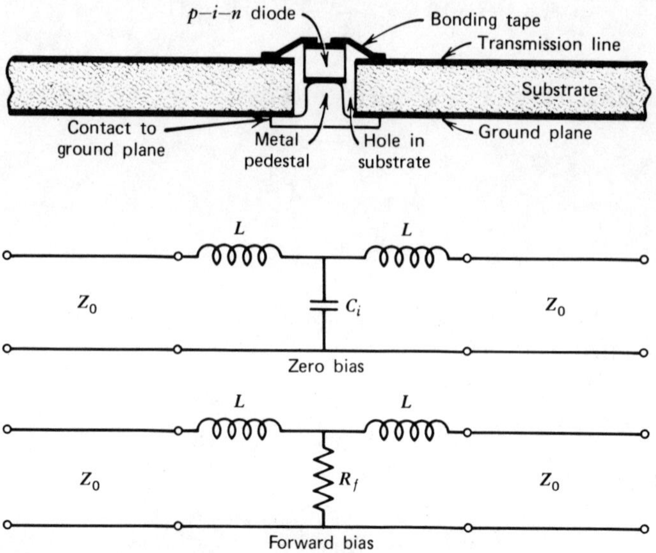

Figure 12.6. Shunt-mounted unpackaged diode on microstrip line ($C_p = L_p = R_s = 0$).

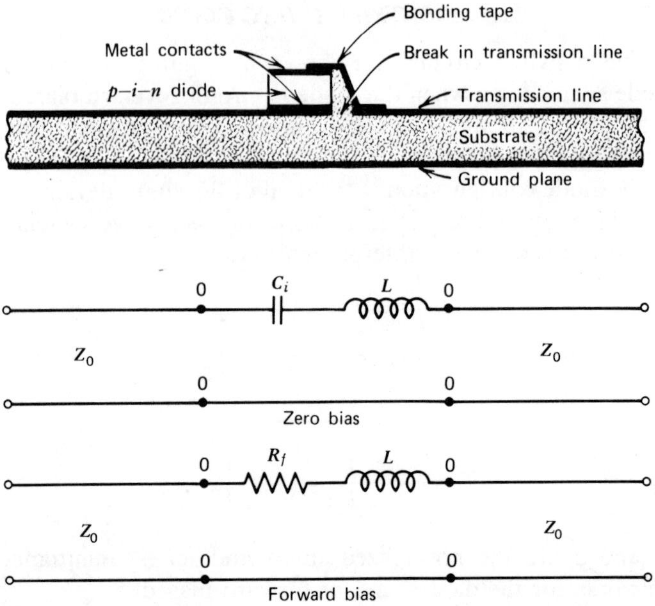

Figure 12.7. Series-mounted unpackaged diode on microstrip line ($C_p = L_p = R_s = 0$).

whereas for the series network it is

$$L(\mathrm{dB}) = 10 \log_{10} \frac{1}{|S_{21}|^2} = 10 \log_{10} \left| 1 + \frac{z}{2} \right|^2 \tag{12.4}$$

The respective reflection coefficients are

$$S_{11} = \frac{-y}{y+2} \tag{12.5}$$

$$S_{11} = \frac{z}{z+2} \tag{12.6}$$

In terms of the original variables, the isolation and insertion loss of the shunt-mounted diode are

$$L(\mathrm{dB}) = 10 \log_{10} \left[1 + \frac{Z_0}{2R_f} \right]^2 \tag{12.7}$$

$$L(\mathrm{dB}) = 10 \log_{10} \left[1 + \left(\frac{\omega C_i Z_0}{2} \right)^2 \right] \tag{12.8}$$

whereas for the series network they are

$$L(\mathrm{dB}) = 10 \log_{10}\left[1 + \left(\frac{1}{2\omega C_i Z_0}\right)^2\right] \tag{12.9}$$

$$L(\mathrm{dB}) = 10 \log_{10}\left[1 + \frac{R_f}{2Z_0}\right]^2 \tag{12.10}$$

In obtaining these results the parasitic inductance and capacitance of the diode package have been omitted in the equivalent circuits in Figure 12.5, which would apply to a packageless diode. Furthermore, it has been assumed that the diode is a low-loss capacitor for reverse bias and a low-value resistor in forward bias. However, the more complete equations can be obtained by straightforward algebra.

The behavior of the shunt diode is depicted in Figure 12.8*a* and 12.8*b*, whereas that of the series diode is shown in Figures 12.9*a* and 12.9*b*. It is

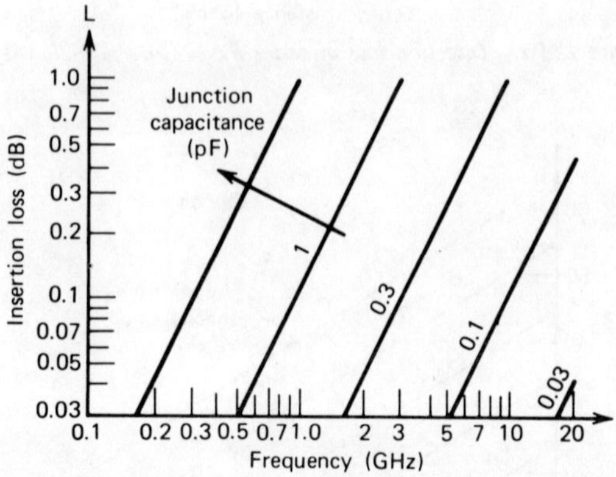

Figure 12.8a. Insertion loss of shunt diode (after P. Chorney).

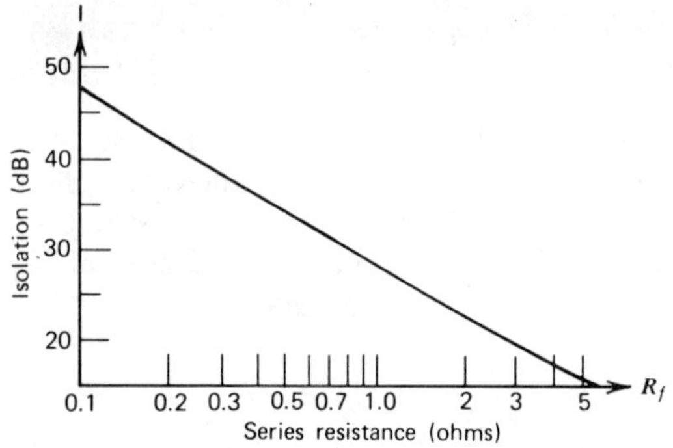

Figure 12.8b. Isolation of shunt diode (after P. Chorney).

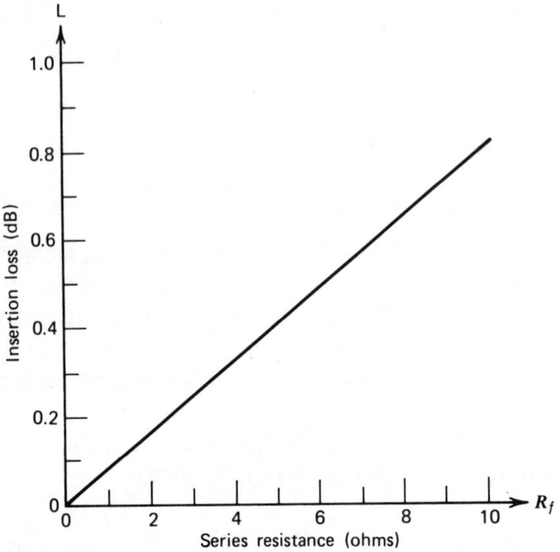

Figure 12.9a. Insertion loss of series diode (after P. Chorney).

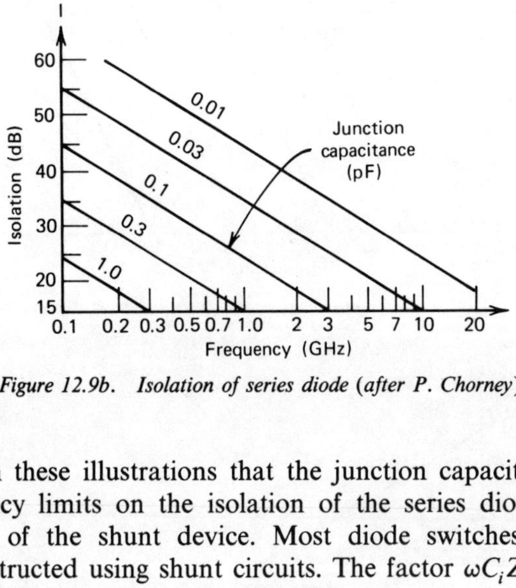

Figure 12.9b. Isolation of series diode (after P. Chorney).

apparent from these illustrations that the junction capacitance C_i places upper frequency limits on the isolation of the series diode and on the insertion loss of the shunt device. Most diode switches are therefore normally constructed using shunt circuits. The factor $\omega C_i Z_0$ is sometimes taken as the figure of merit of the PIN diode.

12.2 *DISSIPATION IN PIN DIODES*

Although diode switches are considered reflective devices, not all the isolation is caused by mismatch loss. A significant portion of the attenuation is resistive loss due to the power absorbed by the device. The dissipation in the diode circuit may be obtained in terms of S_{11} and S_{21} by using the principle of the conservation of energy

$$P_{\text{diss}} = \frac{1}{2} \sum_{i=1}^{2} a_i a_i^* - b_i b_i^* \tag{12.11}$$

This equation states that the power incident on the device must be equal to the power absorbed in the circuit plus the power emerging from the junction at all ports.

The input and output waves are given in the usual way by

$$\begin{bmatrix} b_1 \\ b_2 \end{bmatrix} = \begin{bmatrix} S_{11} & S_{21} \\ S_{21} & S_{11} \end{bmatrix} \begin{bmatrix} a_1 \\ a_2 \end{bmatrix} \tag{12.12}$$

subject to the boundary conditions

$$a_1 = 1 \tag{12.13}$$

$$a_2 = 0 \tag{12.14}$$

The result is

$$b_1 = S_{11} \tag{12.15}$$

$$b_2 = S_{21} \tag{12.16}$$

Substituting these a's and b's into the energy equation gives

$$Q_{11} = 1 - |S_{11}|^2 - |S_{21}|^2 \tag{12.17}$$

which is the result obtained in Chapter 1.

Assuming a pure resistance the result is

$$Q_{11}\,(\text{dB}) = 10 \log_{10} \left[\frac{2\sqrt{g}}{g+2} \right]^2 \tag{12.18}$$

for the shunt diode arrangement and

$$Q_{11}\,(\text{dB}) = 10 \log_{10} \left[\frac{2\sqrt{r}}{r+2} \right]^2 \tag{12.19}$$

for the series diode configuration.

12.3 *NONREFLECTIVE PIN DIODE ATTENUATORS*

Reflective PIN diode attenuators may be combined with symmetrical 3-dB hybrid couplers to obtain nonreflective attenuators. Figure 12.10 indicates a typical arrangement. The operation of this device may be understood by examining its input and output waves with the diodes unbiased and forward biased.

If the diodes are biased they appear as short circuits and the equivalent diagram in Figure 12.11 applies. The input waves for this circuit are

$$a_1 = 1 \tag{12.20}$$

$$a_2 = 0 \tag{12.21}$$

$$a_3 = -b_3 \tag{12.22}$$

$$a_4 = -b_4 \tag{12.23}$$

The outgoing waves at the 4-ports may now be obtained by constructing the input/output relation of the circuit by making use of the scattering matrix of a 3-dB hybrid.

$$\begin{bmatrix} b_1 \\ b_2 \\ b_3 \\ b_4 \end{bmatrix} = \frac{1}{\sqrt{2}} \begin{bmatrix} 0 & 0 & 1 & j \\ 0 & 0 & j & 1 \\ 1 & j & 0 & 0 \\ j & 1 & 0 & 0 \end{bmatrix} \begin{bmatrix} 1 \\ 0 \\ -b_3 \\ -b_4 \end{bmatrix} \tag{12.24}$$

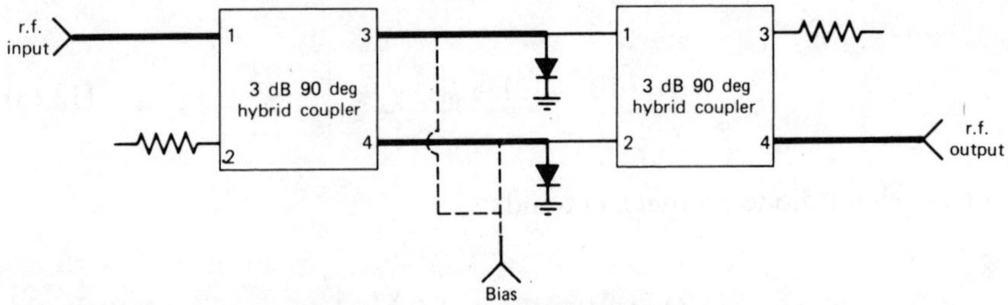

Figure 12.10. Nonreflective PIN diode attenuator.

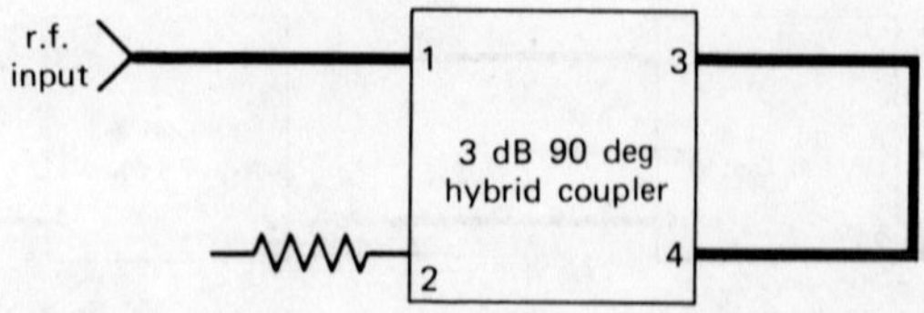

Figure 12.11. Equivalent circuit of PIN diode attenuator with diodes shortcuited.

The result is

$$b_1 = -\frac{1}{\sqrt{2}}(b_3 + jb_4) \tag{12.25}$$

$$b_2 = -\frac{1}{\sqrt{2}}(jb_3 + b_4) \tag{12.26}$$

$$b_3 = \frac{1}{\sqrt{2}} \tag{12.27}$$

$$b_4 = \frac{j}{\sqrt{2}} \tag{12.28}$$

Substituting the last two equations into the first two gives

$$b_1 = 0 \tag{12.29}$$

$$b_2 = -j \tag{12.30}$$

Thus the signal port remains matched while the two reflected waves at ports 3 and 4 combine at the usually decoupled port 2.

If the PIN diodes are unbiased they appear as opencircuits at ports 3 and 4 of the first hybrid. The two hybrids are now in cascade and the equivalent circuit in Figure 12.12 applies. To obtain the input/output relation of this circuit the output waves of the first hybrid are taken as the input waves for the second. The output waves of the second hybrid are

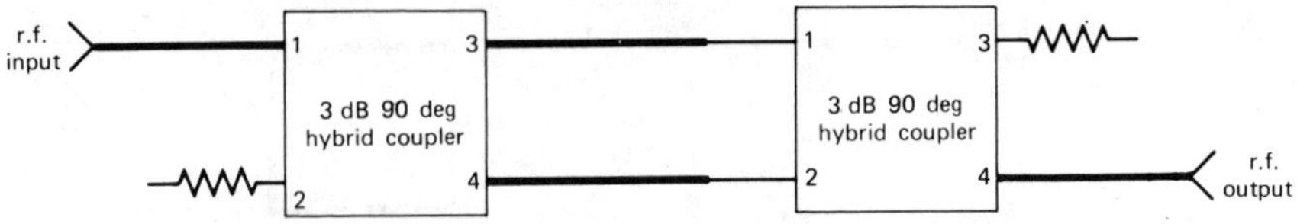

Figure 12.12. Equivalent circuit of PIN diode attenuator with diodes opencircuited.

therefore obtained from the following matrix equation

$$\begin{bmatrix} b_1 \\ b_2 \\ b_3 \\ b_4 \end{bmatrix} = \frac{1}{\sqrt{2}} \begin{bmatrix} 0 & 0 & 1 & j \\ 0 & 0 & j & 1 \\ 1 & j & 0 & 0 \\ j & 1 & 0 & 0 \end{bmatrix} \begin{bmatrix} \frac{1}{\sqrt{2}} \\ \frac{j}{\sqrt{2}} \\ 0 \\ 0 \end{bmatrix} \tag{12.31}$$

Thus

$$b_1 = 0 \tag{12.32}$$

$$b_2 = 0 \tag{12.33}$$

$$b_3 = 0 \tag{12.34}$$

$$b_4 = j \tag{12.35}$$

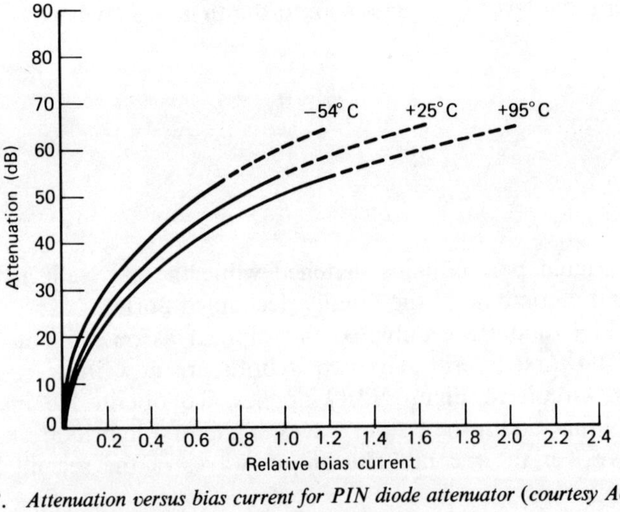

Figure 12.13. Attenuation versus bias current for PIN diode attenuator (courtesy Aertech, Inc.).

so that a wave entering port 1 of the two cascaded hybrids is completely transferred to port 4, none entering the other two ports.

The circuit can therefore be used to vary a wave at port 4 from port 1 by switching the diodes between their biased and unbiased states, the input port remaining matched at all times, and any reflections at the junction of the two hybrids being absorbed in the matched termination at port 2. Figure 12.13 depicts a typical diode current versus attenuation of such a variable attenuator.

12.4 *SINGLE-POLE DOUBLE-THROW PIN DIODE SWITCH USING SHUNT DIODES*

Single pole double throw (SPDT) PIN diode switches may be constructed by mounting diodes in series, shunt, or a combination of the two in two of the three transmission lines of a T or Y junction. Schematic diagrams of the three configurations are given in Figure 12.14. The location of the diodes for the shunt device may be obtained by using the appropriate scattering matrix subject to the boundary conditions that with one diode shortcircuited and the other opencircuited there is no reflection in the input line and only two of the lines of the junction are connected.

For simplicity the following scattering matrix which applies to a compensated junction is employed

$$\bar{S}=\begin{bmatrix} -1/3 & 2/3 & 2/3 \\ 2/3 & -1/3 & 2/3 \\ 2/3 & 2/3 & -1/3 \end{bmatrix} \tag{12.36}$$

The incident waves at the three ports of the junction with an opencircuit at port 2 are

$$a_1=1 \tag{12.37}$$

$$a_2=b_2 \tag{12.38}$$

$$a_3=0 \tag{12.39}$$

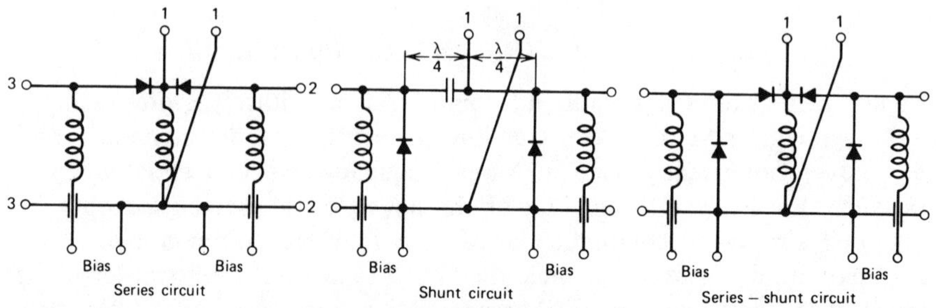

Figure 12.14. SPDT circuits.

Hence

$$\begin{bmatrix} b_1 \\ b_2 \\ b_3 \end{bmatrix} = \begin{bmatrix} -1/3 & 2/3 & 2/3 \\ 2/3 & -1/3 & 2/3 \\ 2/3 & 2/3 & -1/3 \end{bmatrix} \begin{bmatrix} 1 \\ b_2 \\ 0 \end{bmatrix} \tag{12.40}$$

Expanding this equation gives

$$b_1 = -\frac{1}{3} + \frac{2b_2}{3} \tag{12.41}$$

$$b_2 = \frac{2}{3} - \frac{b_2}{3} \tag{12.42}$$

$$b_3 = \frac{2}{3} + \frac{2b_2}{3} \tag{12.43}$$

which leads to

$$b_1 = 0 \tag{12.44}$$

$$b_2 = \tfrac{1}{2} \tag{12.45}$$

$$b_3 = 1 \tag{12.46}$$

This result shows that when an opencircuit is placed at the terminals of port 2 the incident wave at port 1 is transferred to port 3 with no reflection at port 1.

Since, in practice, it is difficult to obtain a perfect opencircuit, the diodes are mounted a quarter-wave away from the junction and are shortcircuited there. The usual application for a single-pole, double-throw switch is digital, where a single input is rapidly switched between two outputs. Since each arm is independently biased, two drivers are required.

12.5 *BASIC CONSTRUCTION OF SP3T SWITCH*

Figure 12.15 shows a broadband commercial multithrow switch using a hybrid integrated-circuit construction. It employs some materials and techniques not usually used in lower frequency hybrid technology. For instance, the leads into and out of the active diode elements cannot be regarded simply as conductors as is true in more common hybrid integrated circuits. The leads, since they have significant electrical length at microwave frequencies, are TEM transmission lines matched in character-

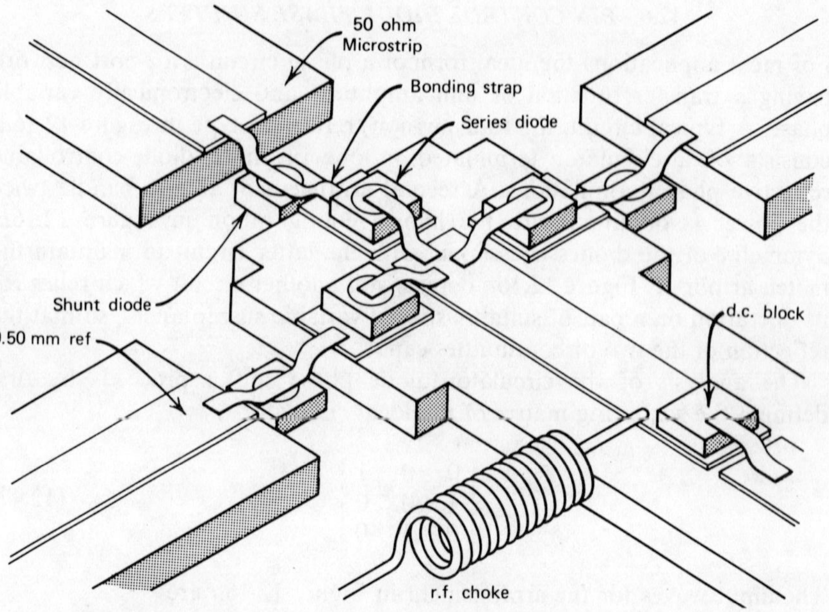

Figure 12.15. Microstrip construction of SP3T switch (after P. Chorney).

istic impedance to the intended application. The microstrip transmission line is "printed" on a teflon-fiberglass substrate clad with copper. Shunt diodes in chip form are shown bonded to the ground plane and series diodes, also in chip form, are shown bonded to the center conductor. The top contacts of the diodes are bonded to metal straps which connect them to the transmission line.

Since parasitic reactances cannot be totally ignored, the straps act as an inductance connected to the diodes. Choosing each strap length and width such that their inductances, together with the associated junction capacitances of the diodes, form lowpass filters eliminates the effect of these parasitic elements at frequencies below the cut-off frequency of the lowpass filter. Consequently, the cut-off frequency of the filters is designed to be above the operating band of interest. The characteristic impedance of the filters is chosen to be approximately 50 ohms.

The d.c. blocking and r.f. bypass capacitors are high-dielectric ceramic chips which are bonded and attached with straps, as are the diode chips. The r.f. chokes are miniature coils wound with enameled wire. The sizes in Figure 12.15 are approximately to scale. The width of the microstrip line is about 0.5 mils, as shown for reference, whereas the diodes are usually 0.4-mil squares about 0.1 mils high.

12.6 *PIN CONTROL DIODE PHASE SHIFTERS*

For most applications the ideal form of a phase circuit is a 2-port network having a transfer function of unit amplitude and electronically variable phase. A typical circuit, the transmission/reflection type in Figure 12.16*a*, consists of a circulator terminated in one or more diode-controllable reflective phase terminations. A second arrangement able to handle twice the power is obtained with the hybrid configuration in Figure 12.16*b*. Symmetry of the diodes is essential with the latter circuit to maintain the match at port 1. Figure 12.16*c* depicts still another circuit which relies for its operation on a pair of suitably spaced variable susceptances, so that the reflection of the two discontinuities cancel.

The analysis of the circulator/diode phase shifter proceeds by first defining the scattering matrix of the ideal circulator

$$\begin{bmatrix} 0 & 0 & 1 \\ 1 & 0 & 0 \\ 0 & 1 & 0 \end{bmatrix} \tag{12.47}$$

The input waves for the arrangement in Figure 12.16*a* are

$$a_1 = 1 \tag{12.48}$$

$$a_2 = b_2 e^{-j2\theta_1} \tag{12.49}$$

$$a_3 = 0 \tag{12.50}$$

where θ_1 is the phase angle of the shortcircuited transmission line at port 2 with the diode on.

The reflected waves are therefore

$$\begin{bmatrix} b_1 \\ b_2 \\ b_3 \end{bmatrix} \begin{matrix} = \\ = \\ = \end{matrix} \begin{bmatrix} 0 & 0 & 1 \\ 1 & 0 & 0 \\ 0 & 1 & 0 \end{bmatrix} \begin{bmatrix} 1 \\ b_2 e^{-j2\theta_1} \\ 0 \end{bmatrix} \tag{12.51}$$

Expanding the above equation gives

$$b_1 = 0 \tag{12.52}$$

$$b_2 = 1 \tag{12.53}$$

$$b_3 = b_2 e^{-j2\theta_1} \tag{12.54}$$

This gives the outgoing wave b_3 as

$$b_3 = e^{-j2\theta_1} \tag{12.55}$$

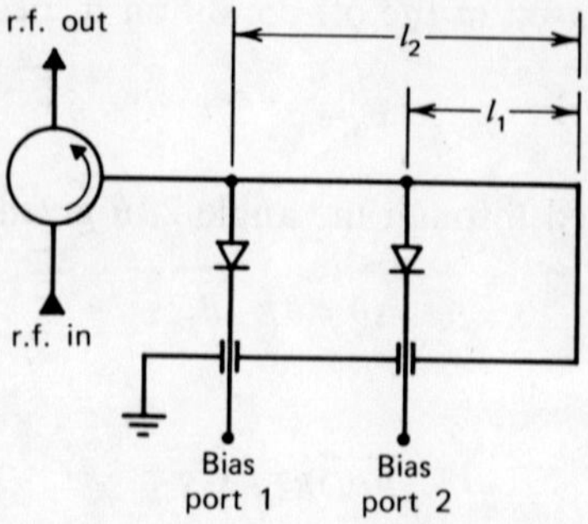

Figure 12.16a. Circulator coupled phase-shifter.

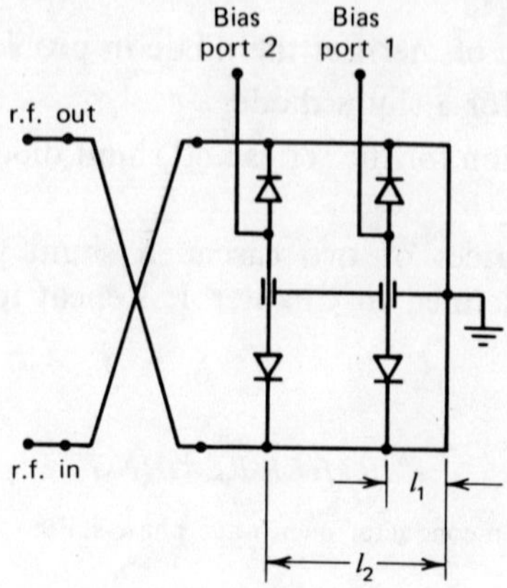

Figure 12.16b. Hybrid coupled phase-shifter.

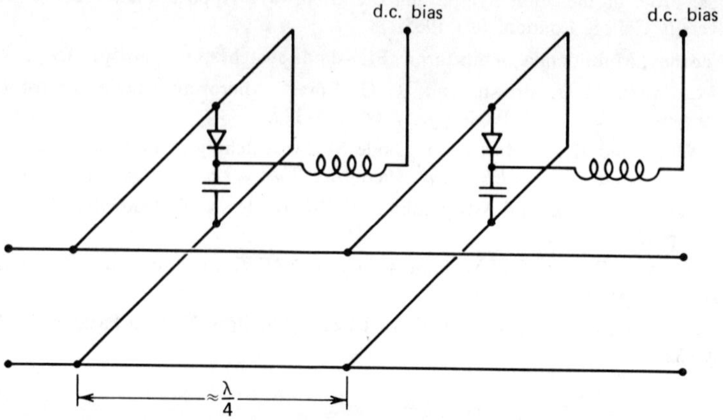

Figure 12.16c. Periodically loaded line phase-shifter.

If the diode is now biased in the off condition b_3 becomes

$$b_3 = e^{-j2\theta_2} \tag{12.56}$$

Thus b_3 is phase shifted through the angle $2\Delta\theta$ given by

$$\Delta\theta = \theta_1 - \theta_2 \tag{12.57}$$

PROBLEMS

1. A series pin diode has a capacitance of 0.1 μF and a forward resistance of 1 Ω. Calculate its insertion loss and attenuation at 1 GHz through 20 GHz in 1 GHz steps.
2. Calculate the figure of merit of the diode in problem 1.
3. Repeat problem 1 for a shunt diode.
4. Obtain the dissipation for the series and shunt diodes in problems 1 and 2.
5. Obtain the $\overline{S}$ matrices of two cascaded shunt pin diodes using the transfer matrices defined in Chapter 1. Repeat for the series arrangement.

FURTHER READING

1. J. White, Review of semiconductor microwave phase-shifters, *Proc. IEEE*, **56**, 11 (Nov. 1968), pp. 1924–1931.
2. J. K. Hunton and A. G. Ryals, Microwave variable attenuators and modulators using *p-i-n* diodes, *Trans. IRE Microwave Theory Tech.* **10** (July 1962), p. 262.
3. L. J. J. Hinton and D. F. Burry, *p-i-n* diode modulators for the *K* and *Q* frequency bands, Proc. of the Joint Symposium on Microwave Applications of Semiconductors University College London, July 1965.
4. P. Chorney, Multi-octave, multi-throw, PIN-diode switches, *Microwave J.* (Sept. 1974).
5. R. M. Ryder, N. J. Brown, and R. G. Forest, Microwave diode control devices, *Microwave J.* (Feb., Mar. 1968), pp. 57–64, 115–122.
6. J. F. White and K. E. Mortenson, Diode SPDT switching at high power with octave microwave bandwidth, *IEEE Trans. Microwave Theory Tech.* **16** (Jan. 1968), pp. 30–36.
7. R. Tenenholtz, Broadband MIC multithrow PIN-diode switches, *Microwave J.* **16** (July 1973), pp. 25–30.
8. R. V. Garver, Theory of TEM diode switching, *IRE Trans. Microwave Theory Tech.* **11** (May 1961), pp. 224–238.
9. Selection and use of microwave diode switches and limiters, Hewlett Packard Application Note 932.

CHAPTER THIRTEEN

Microwave Mixers

G. P. RIBLET,* and G. LO†

A mixer is a common microwave component that combines signals of two different frequencies ω_1 and ω_2 at a nonlinear diode. The difference frequency $\omega_1 - \omega_2$ so generated is extracted. The signal at frequency ω_1 incident at port 3 is from a local oscillator. The signal incident at port 1 at frequency ω_2 is usually the signal from an antenna. The intermediate (i.f.) frequency $\omega_1 - \omega_2$ is further amplified by an i.f. amplifier. The i.f. frequency usually lies between 30 and 70 MHz for conventional r.f. systems. The schematic diagram for the mixer is shown in Figure 13.1.

Often information carried by the microwave carrier lies in the audio or video region so that the conversion of a low-level signal from a microwave to intermediate frequency might seem to be an unnecessary complication. However, the low noise amplification of an i.f. signal is simple and inexpensive. Furthermore, the direct detection (or modulation) of an r.f. carrier with a signal introduces considerable extra noise. The approach that introduces the least amount of extraneous noise is frequently to convert the microwave signal to an i.f. signal, to amplify the i.f. signal, and then to detect the audio or video signal after i.f. amplification.

Metal semiconductor junction diodes are used in mixers because of their small input capacitance, small power loss, and favorable noise characteristics. The diodes can be of the point-contact or Schottky barrier type. The point-contact diode consists of a semiconductor chip, usually *n*-type silicon, contacted by a sharply pointed metal whisker that typically is chemi-

*Microwave Development Laboratories, Natick, Massachusetts.
†Communication Research Center, Ottawa, Canada.

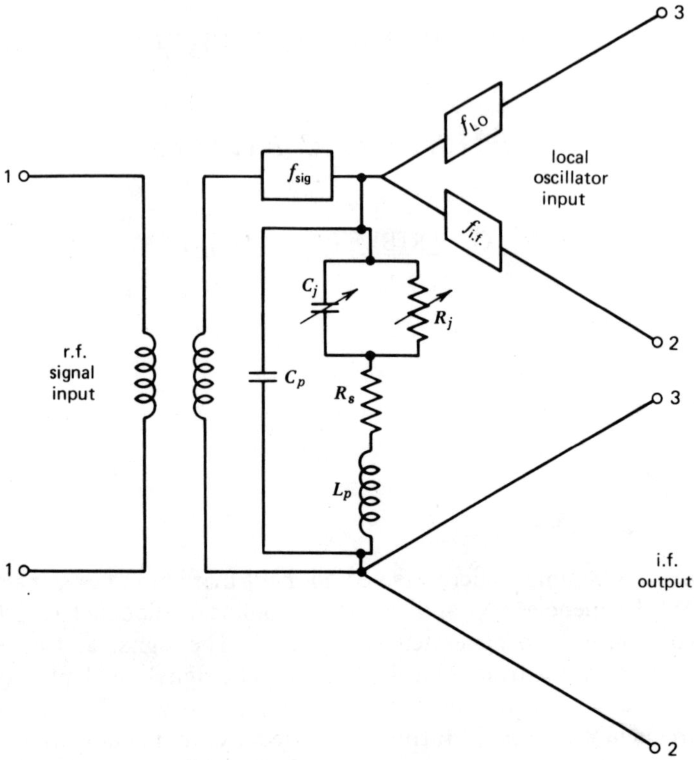

Figure 13.1. Schematic diagram of mixer showing signal, local oscillator, and i.f. terminals.

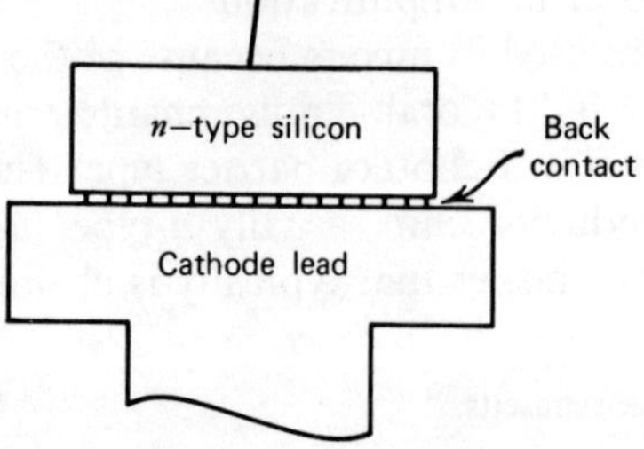

Figure 13.2a. Point contact mixer diode.

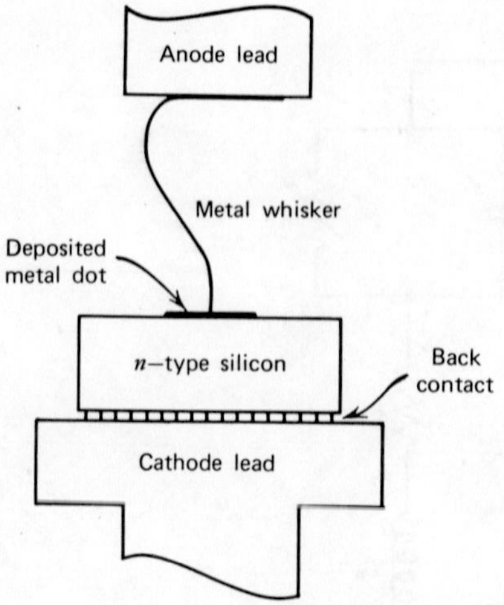

Figure 13.2b. Schottky barrier mixer diode.

Figure 13.3. Construction of Schottky barrier diode.

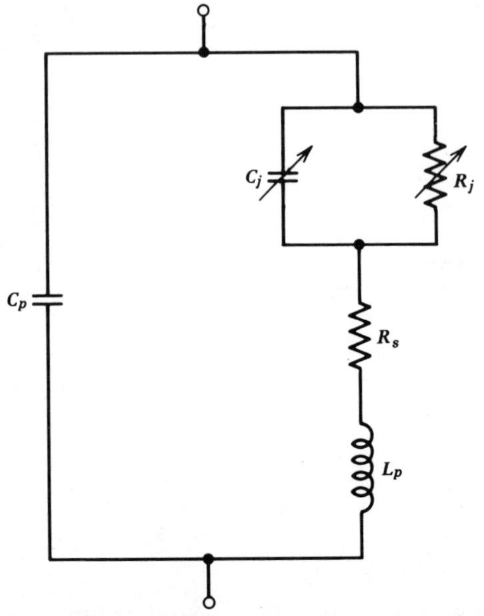

Figure 13.4. Equivalent circuit of mixer diode.

cally etched tungsten wire. The Schottky diode consists of a metal contact in the form of an array of dots deposited on an n-type silicon chip. A metal whisker is used to make contact with one of the deposited metal dots by random probing. This type of construction is uniform and reproducible and the large relative area of planar contact results in a lower series resistance and higher power capability than the point contact diode. Figures 13.2*a* and 13.2*b* show the construction of these two types of diodes, whereas Figure 13.3 depicts a Schottky diode structure. The diode provides a conductance $g(t)$ which is a function of the applied voltage in contrast to a resistor that has a voltage-independent conductance. The equivalent circuit of the mixer is indicated in Figure 13.4. In the metal semiconductor diode both C_j and R_s are treated as constants.

13.1 *SCHOTTKY BARRIER DIODE*

Since in normal applications the local oscillator voltage is much larger than the signal voltage, the latter voltage determines the conductance level g. Furthermore, if the oscillator voltage has a sinusoidal time dependence of the form $\cos\omega_1 t$, then g will be a periodic function of time with period $T=2\pi/\omega_1$. Consequently, $g(t)$ can be expanded in the following Fourier series:

$$g(t) = g_0 + 2g_1 \cos\omega_1 t + 2g_2 \cos 2\omega_1 t + \cdots \tag{13.1}$$

In the case of the Schottky barrier diode, the g_i's may be obtained by starting with the voltage current relation which is fairly accurately given by the following exponential relation

$$i = I_s[\exp(\alpha v) - 1] \tag{13.2}$$

where I_s is the reverse saturation current and α is the exponential coefficient equal to 40 V^{-1} for an ideal diode for which the i-v characteristic is given in Figure 13.5. The incremental conductance of the diode is

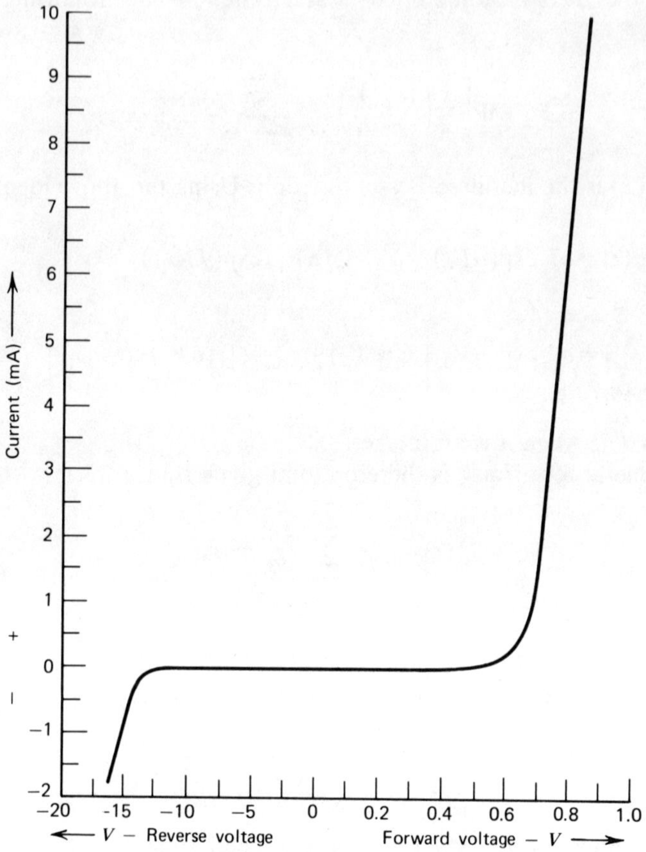

Figure 13.5. i.v. characteristic of Schottky diode.

now defined by

$$g(t)=\frac{di}{dv} \tag{13.3}$$

Taking v in Eq. (13.2) as

$$v=V_0+V_1\cos\omega_1 t \tag{13.4}$$

gives

$$g(t)=\alpha I_s \exp(\alpha V_0)\cdot\exp(\alpha V_1\cos\omega_1 t) \tag{13.5}$$

where V_0 is the bias voltage and $V_1\cos\omega_1 t$ is the local oscillator voltage.

From the theory of modified Bessel functions the following identity applies

$$\exp\left[\frac{x}{2}\left(S+\frac{1}{S}\right)\right]=\sum_{n=-\infty}^{\infty} S^n I_n(x) \tag{13.6}$$

where $I_n(x)$ is the modified Bessel function. Using the above identity gives

$$\begin{aligned} g(t)&=\alpha I_s\exp(\alpha V_0)\sum_{n=-\infty}^{\infty} I_n(\alpha V_1)\exp(jn\omega_1 t)\\ &=\alpha I_s\exp(\alpha V_0)\left[I_0(\alpha V_1)+\sum_{n=1}^{\infty}2I_n(\alpha V_1)\cos n\omega_1 t\right] \end{aligned} \tag{13.7}$$

since $I_n=I_{-n}$ when n is an integer.

The general form for g is therefore that given by Eq. 13.1

$$g(t)=g_0+\sum_{n=1}^{\infty}2g_n\cos n\omega_1 t \tag{13.8}$$

where

$$g_n=\alpha I_s\exp(\alpha V_0)I_n(\alpha V_1) \qquad n=0,1,2\ldots n \tag{13.9}$$

13.2 *RESISTIVE MIXING*

In the previous section it was shown that when a local oscillator voltage is applied to an exponential diode it may be replaced by a time-varying conductance $g(t)$. It will now be shown that the signal current i_s is directly proportional to the product of the signal voltage $V_s \cos\omega_s t$ and the time-varying conductance $g(t)$.

Let the total voltage across the resistance diode junction be

$$v = V_0 + V_1 \cos\omega_1 t + V_s \cos\omega_s t \tag{13.10}$$

Taking a Taylor expansion of the total current about the d.c. and local oscillator voltages and ignoring the terms that do not contain the mixed products,

$$i = V_s \cos\omega_s t \left.\frac{di}{dv}\right|_{v = V_0 + V_1 \cos\omega_1 t} + \tfrac{1}{2} V_s^2 \cos^2\omega_s t \left.\frac{d^2 i}{dv^2}\right|_{v = V_0 + V_1 \cos\omega_1 t} + \cdots \tag{13.11}$$

where $V_s \cos\omega_s t$ is the signal voltage.

Retaining the first term in this series gives

$$i = g(t) V_s \cos\omega_s t \tag{13.12}$$

where $g(t)$ has been defined in Eq. 13.1. Writing $g(t)$ in terms of the original variables yields

$$i = g_0 V_s \cos\omega_s t + \sum_{n=1}^{\infty} 2 g_n V_s \cos n\omega_1 t \cos\omega_s t \tag{13.13}$$

This equation can also be written as

$$i = g_0 V_s \cos\omega_s t + \sum_{n=1}^{\infty} g_n V_s [\cos(n\omega_1 + \omega_s)t + \cos(n\omega_1 - \omega_s)t] \tag{13.14}$$

If down conversion is required the intermediate frequency $n\omega_1 - \omega_s$ is extracted using a suitable filter.

13.3 *CONVERSION LOSS*

The most important parameter of a mixer is the conversion loss between the signal and i.f. terminals. To calculate this parameter it is necessary to obtain a 2-port equivalent circuit for the mixer. Since a mixer contains many frequency products in addition to the desired i.f. some assumption about the termination of the unwanted frequencies is necessary. In the most simple case it is assumed that the terminals of the unwanted frequencies are shortcircuited. If this is the case the total signal impressed across the diode is just the input and i.f. signals v_s and $v_{\text{i.f.}}$

$$v = v_s + v_{\text{i.f.}} \tag{13.15}$$

where

$$v_s = V_s \cos \omega_s t \tag{13.16}$$

and

$$v_{\text{i.f.}} = V_{\text{i.f.}} \cos(\omega_1 - \omega_s)t \tag{13.17}$$

where ω_1 and ω_s are local oscillator and signal radian frequencies.

Retaining the first two terms of $g(t)$ gives the mixer current as

$$i = \left[\left(V_s \cos \omega_s t + V_{\text{i.f.}} \cos(\omega_1 - \omega_s)t\right)\right](g_0 + 2g_1 \cos \omega_1 t) \tag{13.18}$$

where the d.c. and local oscillator components have also been omitted. Separating the signal and i.f. currents and disregarding unwanted side-bands yields

$$i_s = g_0 V_s + g_1 V_{\text{i.f.}} \tag{13.19}$$

$$i_{\text{i.f.}} = g_0 V_{\text{i.f.}} + g_1 V_s \tag{13.20}$$

This result may be described by the following matrix equation

$$\begin{bmatrix} i_s \\ i_{\text{i.f.}} \end{bmatrix} = \begin{bmatrix} g_0 & g_1 \\ g_1 & g_0 \end{bmatrix} \begin{bmatrix} V_s \\ V_{\text{i.f.}} \end{bmatrix} \tag{13.21}$$

Figure 13.6a depicts the 2-port equivalent circuit for the arrangement using the Y-matrix representation. It is identified that $g_0 = Y_{11} = Y_{22}$ and $g_1 = Y_{21} = Y_{12}$.

The required input and output impedances for the mixer can now be derived from the 2×2 $[Y]$ matrix. If it is assumed that an image impedance match exist between input and output, then

$$Y_{\text{in}} = \sqrt{y_{11}\left(y_{11} - \frac{y_{12}^2}{y_{22}}\right)} \tag{13.22}$$

$$Y_{\text{out}} = \sqrt{y_{22}\left(y_{22} - \frac{y_{12}^2}{y_{11}}\right)} \tag{13.23}$$

For image match the signal source admittance is made equal to Y_{in} and the output "if" load admittance is made equal to Y_{out}. Figure 13.6b gives the schematic diagram for this arrangement.
The available power from the signal source is

$$V_1^2 Y_{in} \tag{13.24}$$

The power in the load is

$$V_2^2 Y_{out} \tag{13.25}$$

The conversion loss is now defined as

$$L = \left(\frac{V_1}{V_2}\right)^2 \frac{Y_{in}}{Y_{out}} \tag{13.26}$$

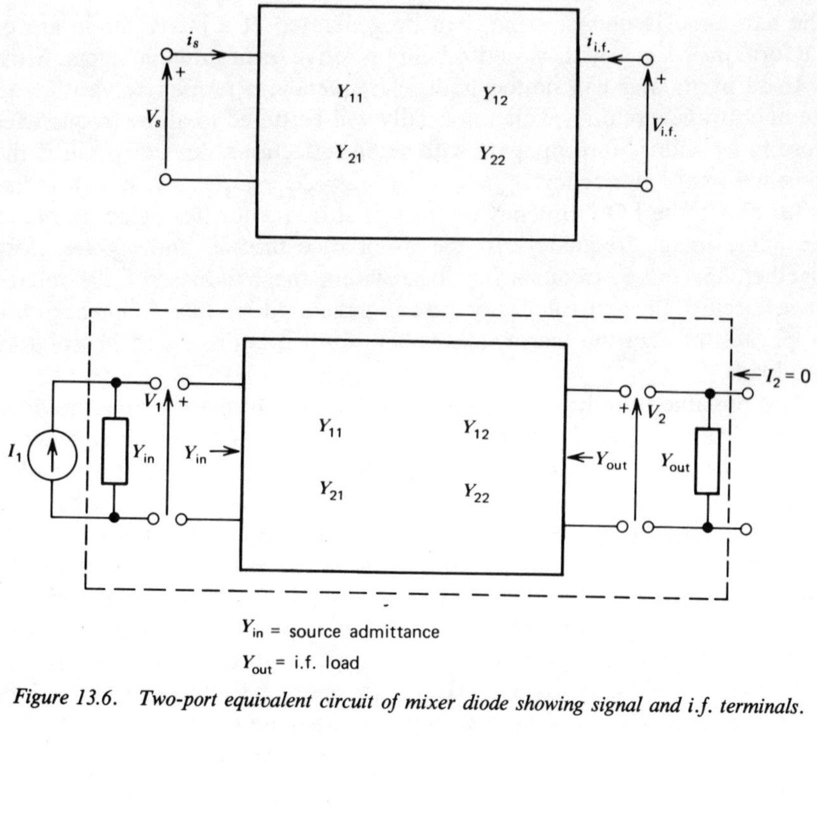

Y_{in} = source admittance

Y_{out} = i.f. load

Figure 13.6. Two-port equivalent circuit of mixer diode showing signal and i.f. terminals.

The voltages V_1 and V_2 can be obtained from the following voltage/current relation by treating the box drawn in dashed lines in Figure 13.6 as a two-port network.

$$\begin{bmatrix} i_1 \\ 0 \end{bmatrix} = \begin{bmatrix} y_{11} + Y_{\text{in}} & y_{12} \\ y_{12} & y_{22} + Y_{\text{out}} \end{bmatrix} \begin{bmatrix} V_1 \\ V_2 \end{bmatrix} \tag{13.27}$$

The result is

$$\left(\frac{V_1}{V_2}\right)^2 = \left(\frac{y_{22} + Y_{\text{out}}}{y_{12}}\right)^2 \tag{13.28}$$

which leads to

$$L = \left(\frac{y_{22} + Y_{\text{out}}}{y_{12}}\right)^2 \frac{Y_{\text{in}}}{Y_{\text{out}}} \tag{13.29}$$

13.4 *IMAGE FREQUENCY*

The harmonic frequencies that can be generated at a mixer diode are of the form $m\omega_1 + n\omega_S$ with m and n being positive or negative integers. Most of these harmonics will lie too high in frequency to propagate very far in the microwave circuitry, which normally will be tuned to allow frequencies close to ω_1 and ω_S to propagate with small reflections. An exception is the so-called image frequency ω_{im} given by $\omega_{\text{im}} = 2\omega_1 - \omega_S (m=2, n=1)$. It lies as far above the LO frequency as the signal frequency lies below it, hence the name image frequency. In the usual case that ω_1 and ω_S are close together, the image frequency will lie within the bandpass of the microwave circuit. In particular any power generated at this frequency will propagate through the microwave system until it is dissipated in isolators and loads.

It is possible to reduce the conversion loss and hence improve the noise figure of some mixers by employing techniques to reflect this image power back to the mixer diodes for further conversion to i.f. Perhaps the simplest scheme for doing this is given in Figure 13.7. A narrow band filter that passes the signal frequency but stops the image frequency is inserted in the input port. The distance of the filter from the mixer diode is varied until a minimum in the conversion loss is obtained. A disadvantage of this technique is that it is only useful for very narrow band applications where ω_S lies within the bandpass of the filter. Furthermore, the insertion loss of the filter, which may be 0.5 dB or so, must be subtracted from the improved conversion loss when computing the noise figure.

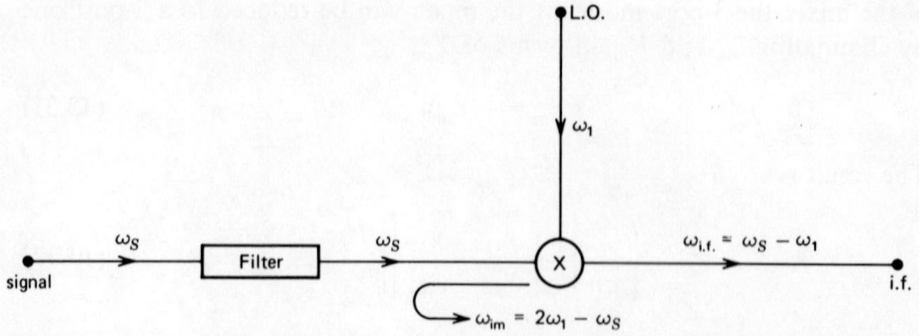

Figure 13.7. Image cancellation mixer.

To account for the image power transmitted to the image frequency position a 3×3 $[Y]$ matrix model for the mixer is required. Assuming that the total signal impressed across the diode is the sum of the signal, i.f., and image voltages, the following 3×3 matrix is obtained using the procedure employed for the 2-port description of the diode.

$$\begin{bmatrix} i_s \\ i_{\text{i.f.}} \\ i_{\text{im}} \end{bmatrix} = \begin{bmatrix} g_0 & g_1 & -g_2 \\ g_1 & g_0 & -g_1 \\ -g_2 & -g_1 & g_0 \end{bmatrix} \begin{bmatrix} V_s \\ V_{\text{i.f.}} \\ V_{\text{im}} \end{bmatrix} \tag{13.30}$$

The equivalent circuit for this arrangement is depicted in Figure 13.8. In the notation used here all currents are assumed to flow towards the diode and all voltages are developed in the same direction.

To study the influence of the image termination on the conversion loss

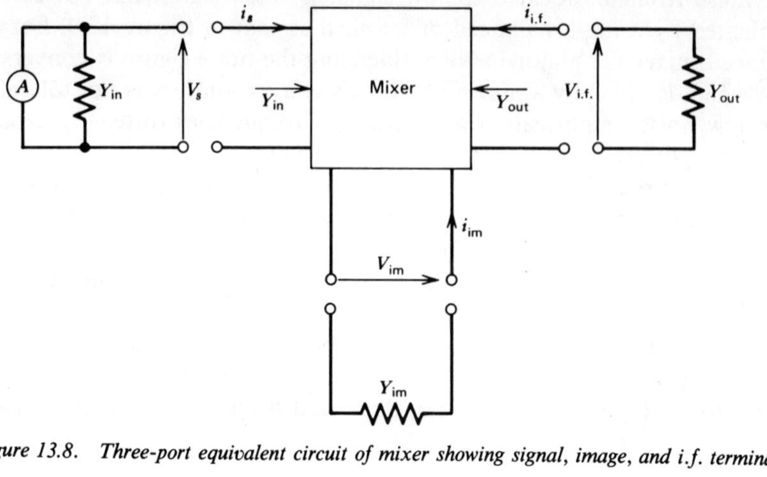

Figure 13.8. Three-port equivalent circuit of mixer showing signal, image, and i.f. terminals.

of the mixer the 3-port model of the mixer can be reduced to a 2-port one by eliminating i_{im} and V_{im} in terms of Y_{im}

$$i_{\text{im}} = -Y_{\text{im}} V_{\text{im}} \tag{13.31}$$

The result is

$$\begin{bmatrix} i_s \\ i_{\text{i.f.}} \end{bmatrix} = \begin{bmatrix} y_{11} & y_{21} \\ y_{12} & y_{22} \end{bmatrix} \begin{bmatrix} V_s \\ V_{\text{i.f.}} \end{bmatrix} \tag{13.32}$$

$$y_{11} = g_0 - \frac{g_2^2}{g_0 + Y_{\text{im}}} \tag{13.33}$$

$$y_{12} = y_{21} = g_1 - \frac{g_1 g_2}{g_0 + Y_{\text{im}}} \tag{13.34}$$

$$y_{22} = g_0 - \frac{g_1^2}{g_0 + Y_{\text{im}}} \tag{13.35}$$

The entries of this 2×2 $[Y]$ matrix may now be directly used in Eq. 13.29 to obtain the conversion loss of the mixer under different image terminations. The usual situations here are $Y_{\text{im}} = o/c$, $Y_{\text{im}} = s/c$, or $Y_{\text{im}} = R_0$. It is observed that the first case is just that dealt with in the previous section. The lowest conversion loss coincides with $Y_{\text{im}} = o/c$, i.e., $Y_{\text{im}} = \infty$.

13.5 *BALANCED MIXER*

The noise from the local oscillator signal is often significant but can be eliminated by using a balanced mixer such as that in Figure 13.9. For the balanced mixer the major factor influencing the noise figure is conversion loss, which is typically about 6 dB. If this conversion loss is not tolerable, then low noise amplification at microwave frequencies (often by a parametric amplifier) must be performed before mixing.

The ring circuit balanced mixer of Figure 13.9 operates as follows. The signal incident at port one splits into 2 waves traveling around the ring circuit in opposite directions. The waves will arrive in phase at both ports 2 and 4, whereas they will be out of phase at port 3 provided the ring sections have the lengths indicated in the diagram. Consequently, ports 1 and 3 will be decoupled. Note that the phases of the signals coupled to ports 2 and 4, respectively, will differ by 180 deg since the path length differs by $\lambda/2$. On the other hand, the signal from the local oscillator will

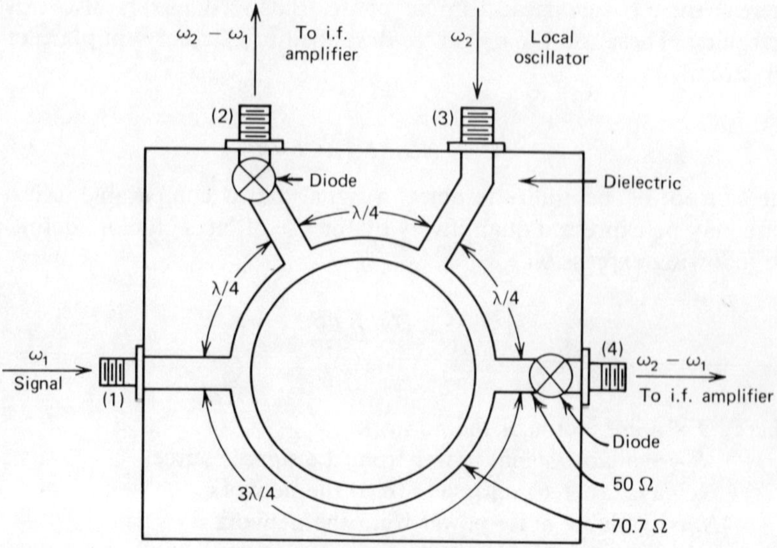

Figure 13.9. Balanced stripline mixer.

couple with the same phase to the diode in port 2 and the diode in port 4. Since ports 1 and 3 are decoupled, the local oscillator signal can't couple into the antenna port. The phase relations described above are essential to understanding the operation of a balanced mixer. Any noise from the local oscillator at the signal frequency will mix with the local oscillator signal to produce i.f. signals at the two diodes that are in phase. The i.f. signals generated by the mixing of the local oscillator and signal frequencies will be out of phase at these two diodes. By feeding the diode outputs into a balanced i.f. input, the local oscillator noise can be made to cancel,

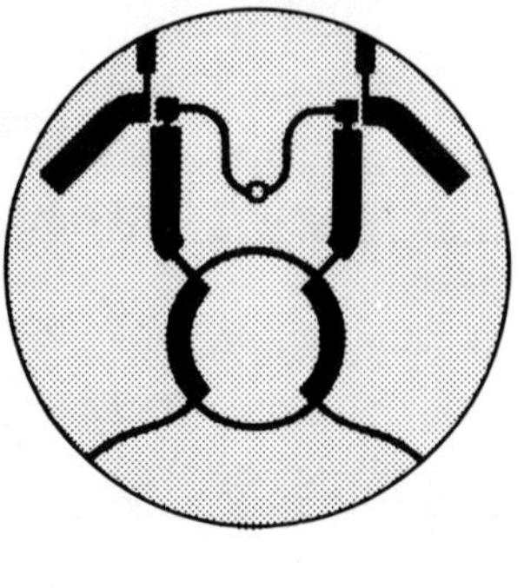

Figure 13.10. Schematic diagram of microstrip balanced mixer (courtesy Mullard).

whereas the i.f. signals add up in phase. Balanced mixers also exist in waveguide. These use the magic tee described in Chapter 4 in place of the ring circuit.

13.6 *NOISE FACTOR*

The concept of the ability to detect a signal that is comparable in level to noise may be expressed quantitively by the use of 'noise factor', defined by the following expression.

$$F=\frac{S/KTB}{S_0/N_0} \tag{13.36}$$

where F=noise factor of the network
S=available signal power from the signal source
S_0=available signal power from the network
N_0=available noise power from the network
K=Boltzmann's constant
T=absolute temperature of the signal source
B=noise bandwidth of the network

The available power gain of the network is defined by

$$G=\frac{S_0}{S} \tag{13.37}$$

The available noise power from the network can be conveniently regarded as arising from a resistor at an equivalent noise temperature T_0, where T_0 is defined by the expression

$$N_0=KT_0B \tag{13.38}$$

Thus Eq. 13.36 may be re-written as

$$F=\frac{1}{G}\frac{T_0}{T} \tag{13.39}$$

Considering the more usual case of two networks in cascade, the noise factor of the combination may be shown to be,

$$F_{12}=F_1+\frac{F_2-1}{G_1} \tag{13.40}$$

where F_1 = is the noise factor of network (1)
F_2 = is the noise factor of network (2)
and G_1 = is the available power gain of network (1)

(Bandwidth of network (2) assumed less than network (1).)
If T_1 is the equivalent noise temperature of network (1), then

$$F_1 = \frac{1}{G_1}\frac{T_0}{T_1} \tag{13.41}$$

or

$$F_1 = \frac{t_1}{G_1} \tag{13.42}$$

where

$$t_1 = \frac{T_0}{T_1} \tag{13.43}$$

Hence expression 13.40 becomes

$$F_{12} = \frac{t_1 + F_2 - 1}{G_1} \tag{13.44}$$

and this form is generally used in the discussion of mixer performance.

The overall noise factor performance of a superheterodyne receiver may be given by

$$F_0 = L_0(t_r + F_{\text{i.f.}} - 1) \tag{13.45}$$

assuming any local oscillator noise is adequately suppressed, where L_0 is the conversion loss of the mixer diode, t_r is the noise temperature ratio of the mixer diode and $F_{\text{i.f.}}$ is the noise factor of the following i.f. amplifier. Hence, L_0 and t_r are the mixer diode parameters determining the value of overall noise factor and these should be minimized.

PROBLEMS

1. Derive the image admittances in Eqs. 13.22 and 13.23 using the schematic diagram in Figure 13.6.
2. Calculate the first three terms of Eq. 13.13 for $\alpha = 40\ V^{-1}$, $I_s = 1\ \mu A$, $V_0 = 0.50\ V$, and $V_1 = 0.10\ V$.

3. Using g_0 and g_1 from problem (2) obtain the conversion loss of the mixer given by Eq. 13.29 for $Y_{\rm im}=o/c$, s/c, and Y_0.
4. Verify that the LO signals are in phase at ports 2 and 4 of the balanced mixer in Figure 13.9.
5. Plot $g(t)$ at 1 MHz retaining g_0, g_1, and g_2.
6. If it is assumed that the parasitic elements C_p, C_j, R_s, and L_p (Figure 13.4) of a mixer diode can be neglected in an analysis, show that the voltage waveform across the diode due to the local oscillator drive is sinusoidal and that $g(t)=1/R_j$. Hence, obtain R_j in terms of g_0, g_1, $g_2 \ldots$, etc.
7. If the parasitic elements in Figure 13.1 are not negligible quantities, the waveform across the diode junction can be periodic but not necessarily sinusoidal. Assuming that a Fourier expansion of the periodic voltage variations across the diode junction can be expressed as

$$v(t)=V_0+\sum_{m=1}^{\infty}(V_m \cos m\omega_l t)$$

derive a new $g(t)$ function based on such a voltage variation across the diode junction.

FURTHER READING

1. H. P. Torrey and C. A. Whitmer, *Crystal Rectifiers*, M.I.T. Radiation Lab. Ser., No. 15. McGraw Hill, 1948.
2. G. C. Southworth, *Principles and Applications of Waveguide Transmission*, Chap. 12, Van Nostrand, New York, 1950.
3. G. C. Messenger and C. T. McCoy, A Low Noise Figure Microwave Crystal Diode, I.R.E. Convention Record, Part 8, 1955, pp. 38–73.
4. T. H. Oxley and J. G. Summers, Metal-Gallium Arsenide Diodes as Mixers, Inst. Phys. Soc. Conf. Ser. No. 3. Gallium Arsenide, 1966.
5. T. H. Oxley and F. Hilsgen, The performance of backward diodes as mixers and detectors at microwave frequencies, *J. I.E.R.E. (Radio and Electronic Engineer)*, **31**, 3 (March 1966) pp. 181–191.

CHAPTER FOURTEEN

Transferred Electron Oscillators and Amplifiers

The Gunn or transferred electron diode is a negative resistance semiconductor device that may be used to construct microwave oscillators and 1-port reflection amplifiers. The origin of the negative resistance is the transfer of electrons from a main conduction energy band to a satellite band in certain semiconductors when the applied electric field in the material exceeds a critical value. A suitable semiconductor material is N-type gallium arsenide provided with two ohmic contacts within a suitable ceramic/metal encapsulation. The construction of a typical Gunn diode is shown in Figure 14.1. The N^+ buffer layers allow simpler ohmic metal contact than is the case with N-type material. The Gunn device is mounted in packages similar to those used in other semiconductor devices such as the varactor and pin diodes. Figure 14.2 illustrates the equivalent circuit of a packaged low-power Gunn diode.

Figures 14.3*a* and 14.3*b* give coaxial and waveguide versions of Gunn oscillators. In its simplest mode the output frequency of *a* transferred electron oscillator (TEO) is determined by a transit time relation, but other enforced modes such as quenched domain modes in resonators may be used to obtain frequency tuning over an octave. Under certain conditions the device may also be operated as an negative resistance amplifier. The schematic in Figure 14.4 illustrates one circuit for a typical 1-port reflection transferred electron amplifier (TEA).

One point to emphasize is that the observed negative differential mobility in TE materials arises purely from bulk effects. Unlike bipolar transistors, tunnel diodes, etc., there are no junctions in TE devices, only ohmic contacts to enable a d.c. voltage bias to be applied across the device, which then produces the required electric field in the bulk material.

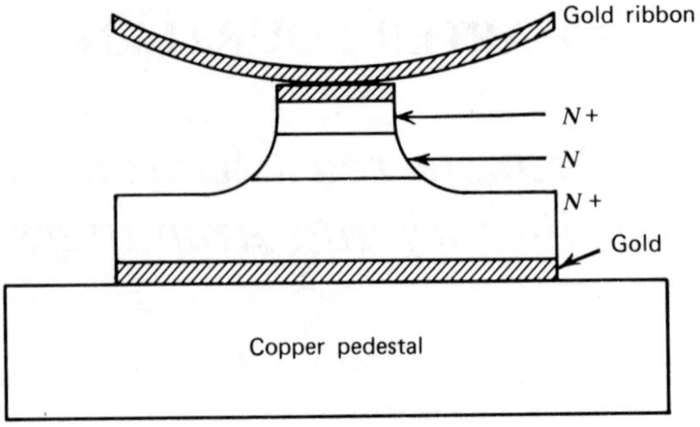

Figure 14.1. Schematic of TE diode.

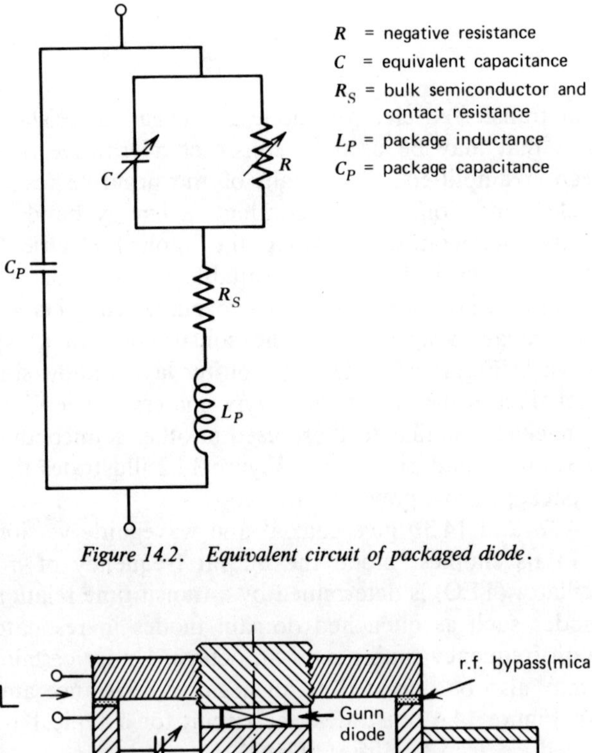

Figure 14.2. Equivalent circuit of packaged diode.

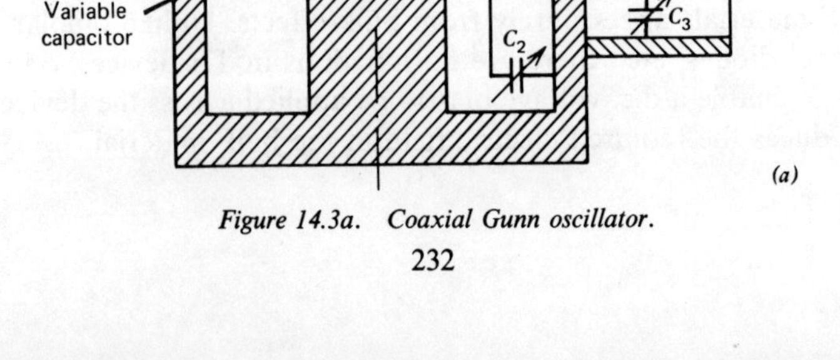

Figure 14.3a. Coaxial Gunn oscillator.

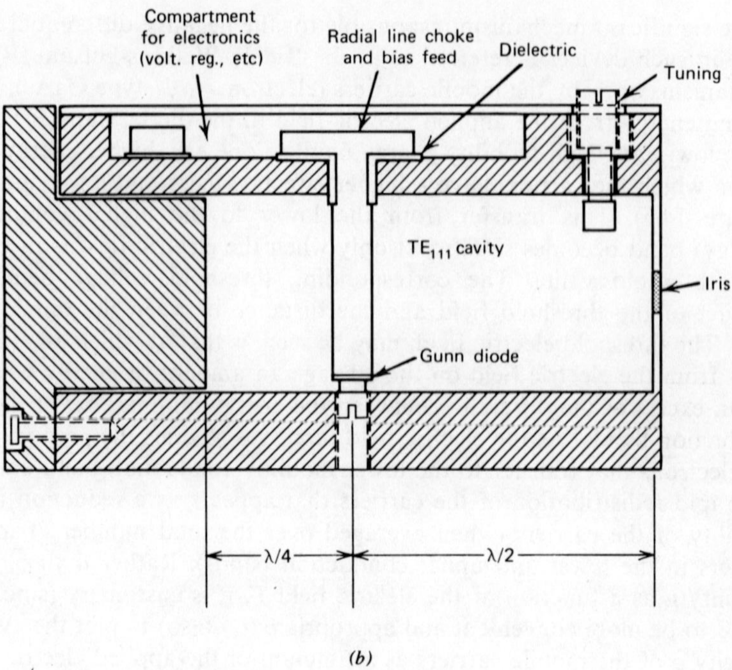

Figure 14.3b. Waveguide Gunn oscillator.

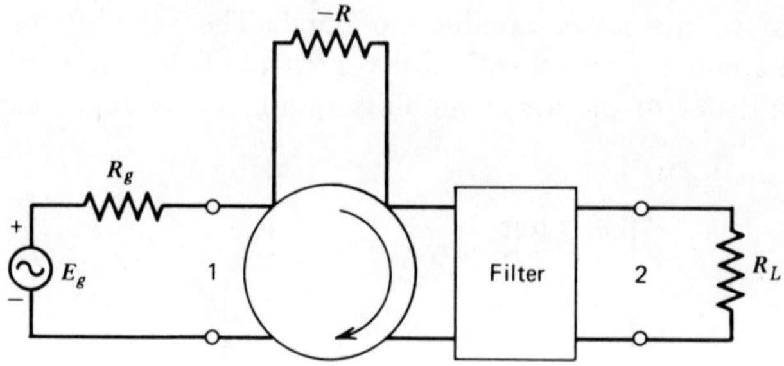

Figure 14.4. Negative-resistance amplifier using circulator.

14.1 *TRANSFERRED ELECTRON TRANSIT TIME DEVICES*

Transferred electron diodes are fabricated by putting ohmic contacts on opposite faces of a slab of *N*-type gallium arsenide (GaAs) and this is then enclosed within a suitable ceramic/metal encapsulation. When the bias voltage applied across the ohmic contacts exceeds a critical value, the device can exhibit a differential negative resistance over a range of frequencies.

The significant mechanism responsible for the negative differential resistance in such devices is referred to as the Ridley-Watkins-Hilsum (RWH) mechanism, whereby the mobile carriers (electrons, in N-type GaAs), upon gaining energy from the applied electric field in the diode, transfer from a main, low-mass, high-mobility energy band to one at a higher energy level and in which the carriers have a higher effective mass and lower mobility (Figure 14.5). This transfer from the lower to the higher conduction (energy) band becomes significant only when the electric field is in excess of a threshold value. The corresponding threshold voltage being the product of the threshold field and the distance between the ohmic contacts. The threshold electric field, may be seen as that at which an electron gains from the electric field on the average an amount of energy which is just in excess of the difference between the energy minima, $\Delta\epsilon$, of the two conduction bands. As the electric field increases so does the proportion of the electrons that transfer to the lower mobility conduction band and it is this rapid redistribution of the carriers that appears as a reduction in the mobility of the carriers (when averaged over the total number of mobile carriers in the lower and upper conduction bands). Rather than plot the mobility μ as a function of the electric field F, it is customary (since it is found to be more convenient and appropriate to do so) to plot the average velocity v of the mobile carriers as a function of the applied electric field, noting that the velocity is the product of the mobility and the electric field ($v = F \cdot \mu$). For a material such as GaAs that exhibits the RWH mechanism the velocity-field plot is typically of the form shown in Figure 14.6.

Below threshold the carriers have a constant, high mobility corresponding to that of the lower conduction band. The velocity thus increases linearly to attain a peak velocity at the threshold field. But then, as carriers begin to transfer to the lower mobility band, the average velocity of the

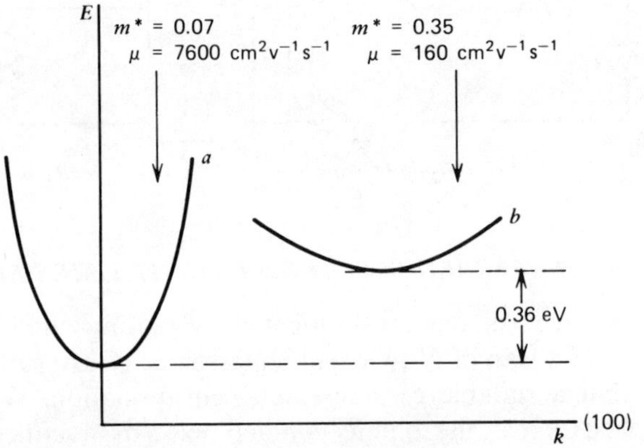

Figure 14.5. Wave vector diagram for conduction band of GaAs.

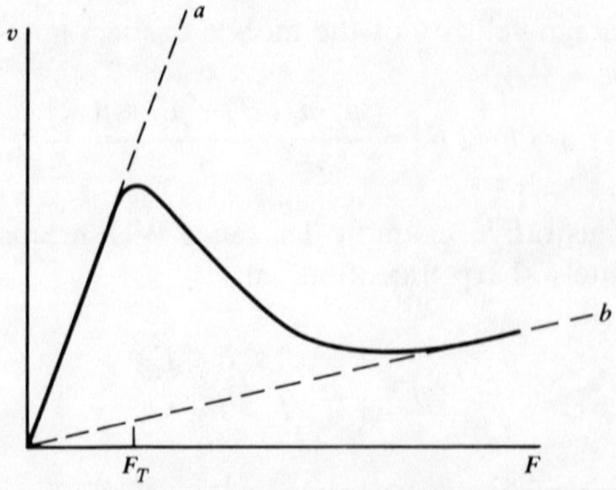

Figure 14.6. Typical velocity field characteristic of Gunn diode.

carriers begin to decrease and continue to do so until all the carriers have transferred from the lower to the upper conduction band at which point the saturated velocity is reached. Beyond the threshold field, and, up to the point where saturated velocity is attained, the material exhibits negative differential mobility (NDM) and is thus the region of greatest interest from the device viewpoint. At electric fields in excess of that corresponding to the saturated velocity, all the carriers are in the upper band and move with the constant low mobility of that band. The velocity-field curve should then have a positive constant slope as shown, but sufficiently reliable experimental results do not exist to corroborate the theoretical predictions.

The above description of the TE mechanism may be put quantitatively in terms of a few simple equations. Thus let the subscript L refer to quantities in the lower, high mobility conduction band, whereas the subscript "U" refers to those in the upper, low mobility conduction band. The total number of mobile carriers n (electrons) in the two conduction bands is a constant.

$$n = n_L(F) + n_U(F)$$

The numbers of mobile carriers in each of the bands are a function of the applied electric field F in the material. (Without any loss of generality the numbers may be taken as those per unit volume, i.e., mobile carrier densities.)

What has been referred to in the text as the (effective) mobility may now be expressed as

$$\mu(F) = \frac{\mu_L \cdot n_L(F) + \mu_U \cdot n_U(F)}{n}$$

and hence the average velocity of the mobile carriers is

$$v(F)=F\cdot\mu(F)=\frac{\mu_L\cdot n_L(F)+\mu_U\cdot n_U(F)}{n}\cdot F$$

As a simple, illustrative example let the RWH mechanism be represented as an infinitely sharp transition, i.e.,

$$n_L(F)=\begin{cases}1 & 0\leqslant F\leqslant F_T\\ 0 & F>F_T\end{cases}$$

and since n is a constant,

$$n_U(F)=\begin{cases}0 & 0\leqslant F<F_T\\ 1 & F>F_T\end{cases}$$

Substituting these in the expression for $v(F)$ at once yields

$$v(F)=\begin{cases}\mu_L\cdot F & 0\leqslant F<F_T\\ \mu_U\cdot F & F>F_T\end{cases}$$

F_T is the threshold field at which, for this ideal case, all carriers move from the lower to the upper band. From the above it is evident that if a TE device is biased to produce a field in excess of the threshold one then the material will exhibit NDM which will appear across the device terminals as negative differential resistance over a range of frequencies.

14.2 *DOMAIN FORMATION AND THE GUNN EFFECT*

In a practical sample of GaAs, the effect of inhomogeneities in the crystal lattice will cause the originally uniform field and carrier distribution to change completely into one that is very nonuniform. The carrier distribution rearranges to form a very thin dipole region, the domain, an accumulation, and depletion layer, and this shape traverses the length of the diode at a velocity roughly equal to the carrier saturated drift velocity.

Outside of this thin dipole region, the carrier density is essentially uniform and so the corresponding electric field pattern is also uniform everywhere (at subthreshold field) except in the domain where it is triangular and above threshold. The domain nucleates from microscopic inhomogeneties at the cathode end of the diode, traverses to the anode end, and then quenches. Within a very short time a new domain forms and starts moving from the cathode end. The current through the diode when the

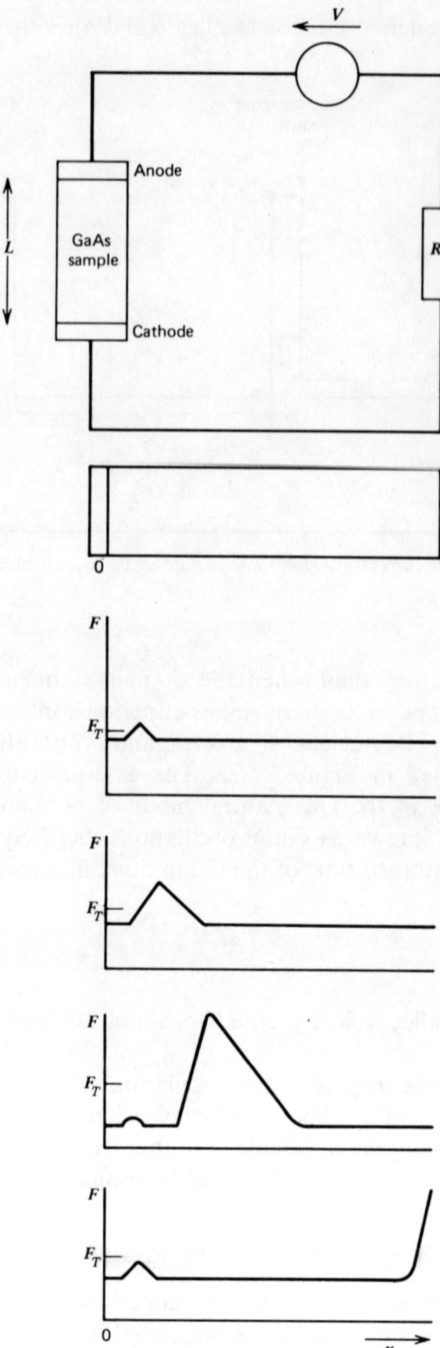

Figure 14.7a. Nucleation, growth, and propagation of a high field domain in n-type GaAs biased beyond the threshold field in a resistive circuit.

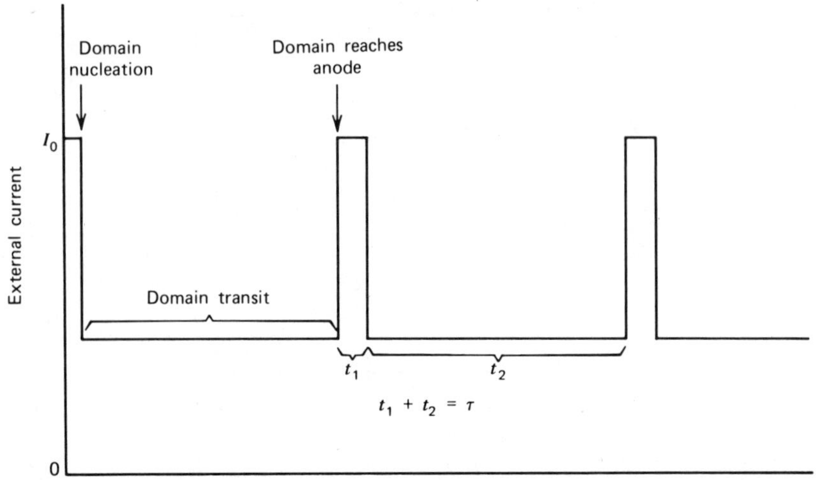

Figure 14.7b. External current waveform appropriate to the basic transit-time mode of operation.

domain exists is lower than when the domain is quenched, and so the above instability appears as sharp spikes at periodic intervals in the current through the diode. The nucleation, growth, and propagation of a high field domain is illustrated in Figure 14.7a. The external current waveform is depicted in Figure 14.7b. This natural mode of oscillation in transferred electron devices is known as Gunn oscillations, the frequency of which is determined by the transit time of the Gunn domain across the diode.

$$T = \frac{L}{v_d}$$

where v_d is the valley velocity equal to about 10^7 cm/sec and T is the transit time.

Although the frequency of Gunn oscillations can be varied slightly by varying the applied bias, almost an octave change in frequency may be obtained by operating the TE diodes in other (forced) modes of operation by placing the device in externally tunable resonant circuits.

14.3 *GAIN OF SIMPLE NEGATIVE RESISTANCE AMPLIFIER*

The schematic diagram of a simple reflection-type negative-resistance amplifier using an ideal circulator is illustrated in Figure 14.4. The gain of such an amplifier may be obtained by replacing G by $-G$ in the definition of the reflection coefficient of the 1-port circuit terminating port 2 of the circulator.

The transducer gain of the circulator terminated in a positive conductance G is

$$|S_{11}(G)|^2 = \left|\frac{Y_0 - G}{Y_0 + G}\right|^2 \tag{14.1}$$

If the positive real termination G is replaced by $-G$ the preceding equation becomes

$$|S_{11}(-G)|^2 = \left|\frac{Y_0 + G}{Y_0 - G}\right|^2 \tag{14.2}$$

Thus, the power reflection coefficient of a negative conductance amplifier is defined as the reciprocal of that associated with a positive real termination.

$$S_{11}(-G) = \frac{1}{S_{11}(G)} \tag{14.3}$$

Equation 14.3 relates the performance of the negative resistance reflection amplifier to that of the same network operating as an impedance matching network terminated in a positive resistance with the same value as that of the negative resistance amplifier. Thus, the problem of finding the maximum gain over a given band is reduced to that of determining the minimum value of S_{11} if the output termination $-G$ is replaced by G. This is just the familiar matching problem of a 1-port circuit.

14.4 *GAIN-BANDWIDTH PRODUCT OF NEGATIVE RESISTANCE AMPLIFIER*

In practice the equivalent circuit of the Gunn device is not the simple negative conductance illustrated in Figure 14.4 but consists of the shunt arrangement of a negative conductance G and a capacitance C in Figure 14.8*a*. It is therefore necessary to locate an inductance across the capacitance to resonate the circuit. Thus the gain of the device is now frequency dependent making it necessary to define a gain-bandwidth product for the device.

The admittance of the tuned negative amplifier in Figure 14.8*b* is

$$Y(\omega) = -G + j\omega_0 C\left(\frac{\omega}{\omega} - \frac{\omega_0}{\omega}\right) \tag{14.4}$$

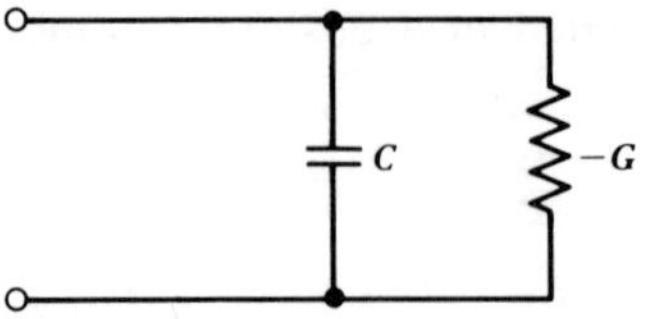

Figure 14.8a. Equivalent circuit of Gunn diode.

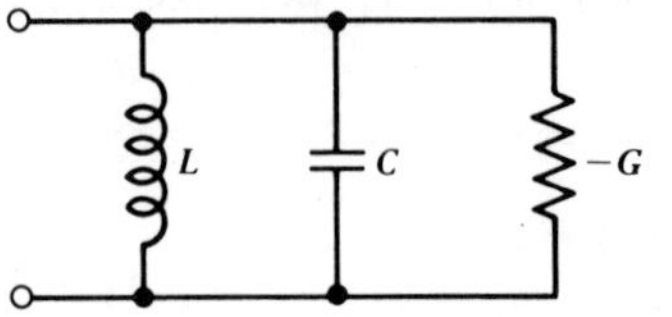

Figure 14.8b. Equivalent circuit of tuned Gunn diode.

The power reflection coefficient is therefore

$$|S_{11}(\omega)|^2 = \frac{(Y_0+G)^2+\omega_0^2C^2\left(\dfrac{\omega}{\omega_0}-\dfrac{\omega_0}{\omega}\right)^2}{(Y_0-G)^2+\omega_0^2C^2\left(\dfrac{\omega}{\omega_0}-\dfrac{\omega_0}{\omega}\right)^2} \tag{14.5}$$

The 3-dB bandwidth is now defined as

$$|S_{11}(\omega_{1,2})|^2 = \frac{|S_{11}(0)|^2}{2} \tag{14.6}$$

where $\omega_{1,2}$ are the two radian frequencies where the gain is 3-dB down. Thus

$$\frac{(Y_0+G)^2+\omega_0^2C^2\left(\dfrac{\omega_{1,2}}{\omega_0}-\dfrac{\omega_0}{\omega_{1,2}}\right)^2}{(Y_0-G)^2+\omega_0^2C^2\left(\dfrac{\omega_{1,2}}{\omega_0}-\dfrac{\omega_0}{\omega_{1,2}}\right)} = \frac{1}{2}\left(\frac{Y_0+G}{Y_0-G}\right)^2 \tag{14.7}$$

Rearranging this equation gives

$$\frac{2G\omega_0}{(\omega_2-\omega_1)\omega_0C} = (|S_{11}(0)|-1)\left(1-\frac{2}{|S_{11}1(0)|^2}\right)^{1/2} \tag{14.8}$$

If $S_{11}(0) \gg 1$ a simplified result is

$$(\omega_2 - \omega_1)|S_{11}(0)| = \frac{2G}{C} \tag{14.9}$$

14.5 *GAIN BANDWIDTH OF WIDEBAND NEGATIVE RESISTANCE AMPLIFIER*

To extend the gain-bandwidth product of the simple tuned amplifier defined by Eqs. 14.8 or 14.9 some form of wideband matching network is necessary. The maximum gain-bandwidth that can be achieved by this means has been defined by Bode by the following integral equation:

$$\int_0^\infty \log_e \left| \frac{1}{S_{11}(j\omega)} \right| d\omega \leqslant \frac{\pi G}{C} \tag{14.10}$$

This integral equation suggests that the maximum bandwidth is obtained with this type of network by having S_{11} a constant over the required frequency interval and unity elsewhere. Assuming this idealized situation

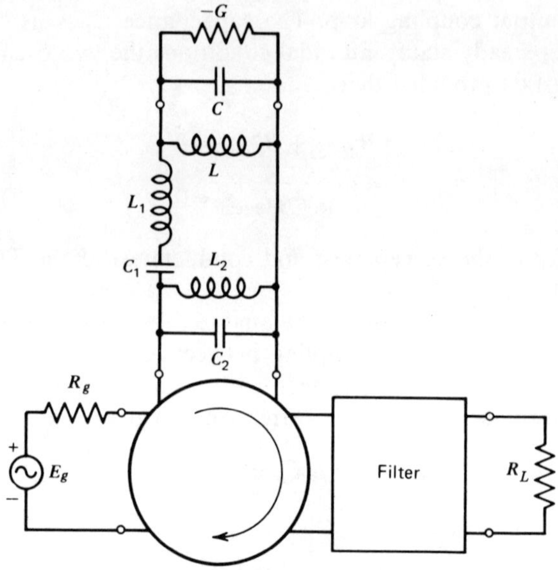

Figure 14.9. Schematic diagram of triple-tuned negative-resistance amplifier.

the above integral becomes

$$(\omega_2-\omega_1)\log_e\frac{1}{S_{11}}=\frac{\pi G}{C} \tag{14.11}$$

For a negative resistance amplifier S_{11} is replaced by $1/S_{11}$ in Eq. 14.11. The result is

$$(\omega_2-\omega_1)\log_e S_{11}=\frac{\pi G}{C} \tag{14.12}$$

This result may be compared with that obtained for a single tuned circuit in Eq. 14.9.

Figure 14.9 depicts a tripled tuned negative resistance amplifier using a 3-port circulator.

14.6 *COAXIAL GUNN OSCILLATOR*

One simple Gunn oscillator circuit that can be readily analyzed is the coaxial arrangement in Figure 14.3*a*. It consists of a coaxial cavity with a TE diode mounted in the reentrant gap between the inner conductor and one end plate of the cavity. The impedance and length of the coaxial cavity determines the frequency of the oscillator, whereas the correct resistive loading on the diode is obtained by adjusting the penetration and orientation of the output coupling loop. The capacitance C_1 is used to tune the device. Under steady-state sinusoidal conditions the two oscillation conditions are satisfied provided that

$$B_{\text{device}}+B_{\text{circuit}}=0 \tag{14.13}$$

$$G_{\text{device}}+G_{\text{circuit}}=0 \tag{14.14}$$

B and G refer to the susceptance and conductance of the TE device and cavity circuit.

The device in Figure 14.3*a* can be analyzed by means of the equivalent circuit in Figure 14.10. The coupling between the loop and the cavity is represented by an ideal transformer with a turns ratio n. Applying the criteria for steady-state operation at frequency ω gives

$$j\omega C-G+j\omega C_1+Y_{s/c}+Y_L=0 \tag{14.15}$$

where

$$Y_{s/c}=-j\cot\left(\frac{2\pi L}{\lambda_0}\right) \tag{14.16}$$

and L is the length of the cavity.

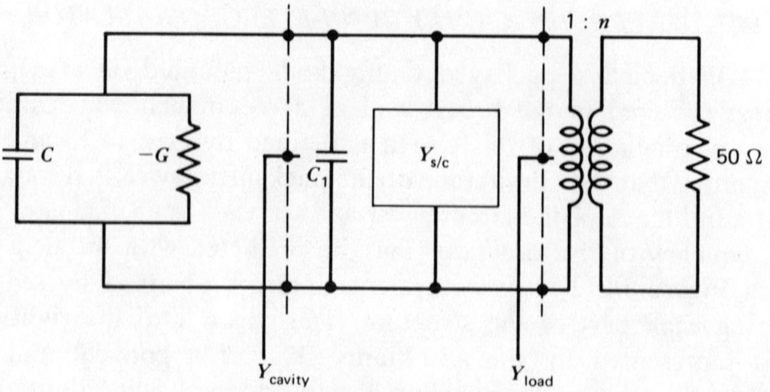

Figure 14.10. Equivalent circuit of coaxial oscillator.

The two oscillation conditions are therefore

$$\omega C+\omega C_1-\cot\left(\frac{2\pi L}{\lambda_0}\right)+\mathrm{Im}(Y_L)=0 \tag{14.17}$$

$$-G+\mathrm{Re}(Y_L)=0 \tag{14.18}$$

Figures 14.11*a* and 14.11*b* depicts a YIG tuned stripline Gunn oscillator and its equivalent circuit.

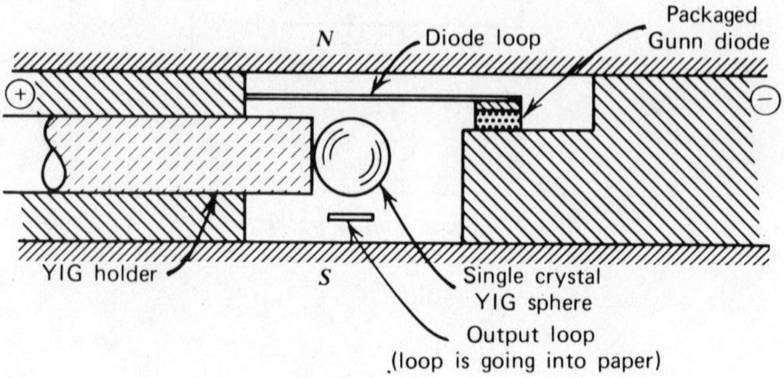

Figure 14.11a. YIG tuned stripline Gunn oscillator.

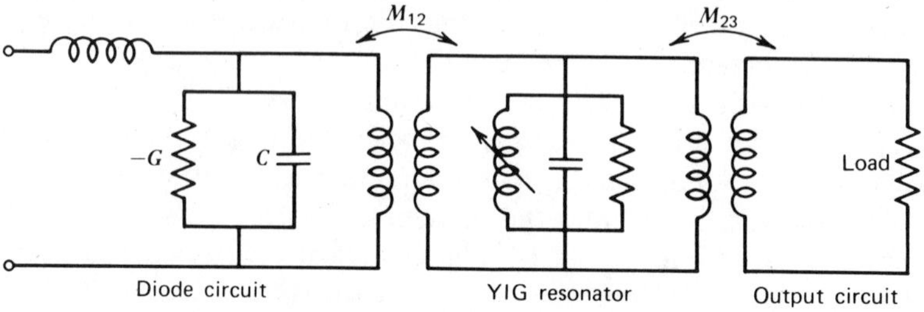

Figure 14.11b. Equivalent circuit of YIG-tuned stripline Gunn oscillator.

14.7 *WAVEGUIDE-CAVITY GUNN-EFFECT OSCILLATOR*

Figure 14.3*b* depicts a packaged Gunn diode mounted on a cylindrical metal post centered in the broad wall of a rectangular waveguide. The frequency of oscillation of this circuit is defined by Eqs. 14.13 and 14.14, which requires that the admittance of the packaged device is resonated by the total admittance of the circuit.

The frequency of this oscillator may be predicted with the help of the circuit in Figure 14.12. This equivalent circuit is obtained by separately considering each part of the structure. The network to the right of the diode is represented by the admittance $Y_{s/c}$ of a conventional short-circuited length of transmission line L with characteristic admittance Y_0 and propagation constant γ.

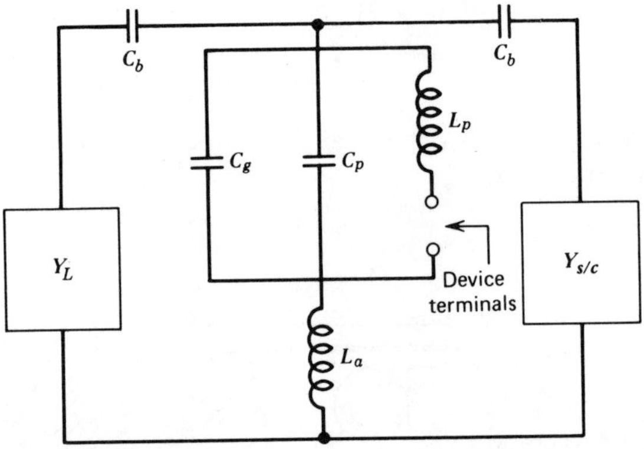

Figure 14.12. Equivalent circuit of post-mounted packaged Gunn waveguide oscillator.

The resonant post can be represented by an equivalent T network composed of series susceptances C_b and a shunt susceptance consisting of the series combination of L_a and C_g. The series capacitors are related to the finite diameter of the post and may be omitted for an infinitely thin post. The inductance L_a is related to the magnetic energy stored in a simple full-height post, whereas the capacitance C_g is related to the capacitive energy stored in the gap between the end of the post and the waveguide. The admittance Y_L is the load admittance transformed to the plane of the post. It includes the matching iris necessary to establish the correct real part G_L in Eq. 14.14. The entries of the packaged diode $-G$, C, R_S, L_p, and C_p having the meaning defined in Figure 14.2. The two oscillation conditions for the circuit in Figure 14.12 is left as an exercise for the reader.

FURTHER READING

1. Wei-Ching Tsai, F. J. Rosenbaum and L. A. MacKenzie, Circuit analysis of waveguide cavity Gunn-effect oscillator, *IEEE Trans. Microwave Theory Tech.*, **18**, 11, (Nov. 1970) pp. 808.
2. C. P. Jethwa and R. L. Gunshor, An analytical equivalent circuit representation for waveguide mounted Gunn oscillators, *IEEE Trans. Microwave Theory Tech.*, **20** (Sept. 1972) pp. 565–572.
3. M. Dean and M. J. Howes, J-Band transferred electron oscillators, *IEEE Trans. Microwave Theory Tech.* **21** (March 1973) pp. 121–127.
4. B. C. Taylor, Microwave semiconductor generation devices, *Electronic Components*, Sept. 1969.
5. A. O. Scanlon and J. G. Lim, A design theory for optimum broadband reflection amplifiers, *IEEE Trans. Microwave Theory Tech.*, **12** (Sept. 1964) pp. 504–511.
6. P. J. Bulman, G. S. Hobson, and B. C. Taylor, *Transferred Electron Devices*, Academic Press, London, 1972.
7. D. V. Morgan and M. J. Howes, *Solid State Electronic Devices*, The Wykeham Science Series, London, 1972.

PROBLEMS

1. A Gunn diode has a negative conductance of 0.002 mhos and a capacitance of 0.30 $\mu\mu$F at 10 GHz. Find the shunt inductance necessary to resonate the circuit.
2. Find the gain and 3-dB bandwidth of the amplifier in problem 1.
3. Obtain the maximum achievable bandwidth for the diode in problem 1.
4. Find the two oscillation conditions for the waveguide oscillator in Figure 14.12 in terms of the real and imaginary parts of Y_L for an unpackaged diode ($C_P = L_P = 0$) and an infinitely thin post ($C_b = 0$).
5. Repeat problem 4 for a packaged diode.
6. Verify Eq. 14.8.
7. Assuming $Y_{s/c}$ in Eq. 14.16 find an expression for L for the waveguide oscillator in Figure 14.12 using an unpackaged diode mounted on an infinitely thin post.
8. A diode has a capacitance C of 0.30 $\mu\mu$F and a negative conductance of 0.002 mhos. Find the turns ratio n and piston location L at 10 GHz for the coaxial oscillator in Figure 14.9. Assume $Y_0 = 0.02$ mhos, $C_1 = 0$, Im-$(Y_L) = 0$.

9. In transferred electron oscillators dipole (Gunn) domains form periodically at the cathode and travel at constant velocity through the length of the semiconductor slab. The current waveform at the terminals of the device is flat at all times except for a very sharp spike when the domain quenches at the anode. The domain moves typically at the valley velocity of the carriers (Figure 14.6), which is about 10^7 cm/sec. For GaAs calculate the length (in microns) of semiconductor (GaAs) material that must be used to make a Gunn oscillator which will yield a fundamental (transit-time) frequency of 10.0 GHz.

CHAPTER FIFTEEN

Microwave Transistor Amplifier Design

(J. HELSZAJN AND W. T. NISBET*)

The transistor, which is basically a three terminal device has one common terminal in forming input and output ports. Depending on which terminal is taken as common, a bipolar transistor is said to be connected in either common base, common emitter or common collector. For a field effect transistor (FET) the transistor is connected in common gate, common source, or common drain. The schematic diagrams for these two types of transistors are shown in Figures 15.1*a* and *b*. Since at microwave frequencies the noise figure of the FET is lower than that of the bipolar transistor only the former is considered in this text. Figure 15.2 shows a MESFET (metal semiconductor field effect transistor).

At microwave frequencies the transistor is universally described by its 2-port scattering matrix since other descriptions require open- and short-circuited measurements that are difficult to establish there. It is found that the scattering matrix is indeed all that is required for the complete design of amplifiers. Tables 15.1 and 15.2 give the S parameters of a transistor under two different bias conditions over a frequency range 1–12 GHz.

One important consideration in the design of amplifiers is the problem of stability. An amplifier may either be unconditionally or conditionally stable. The boundary conditions between these two states is treated in detail as a preamble to the design of amplifiers. It is seen that the complex conjugate image matching of an unconditionally stable transistor leads to the maximum transducer gain for the amplifier.

One important assumption to simplify the design of the amplifier is to assume that the transistor is unilateral, i.e., $S_{12}=0$. Indeed, in practice, it is

*Heriot-Watt University, Edinburgh, Scotland.

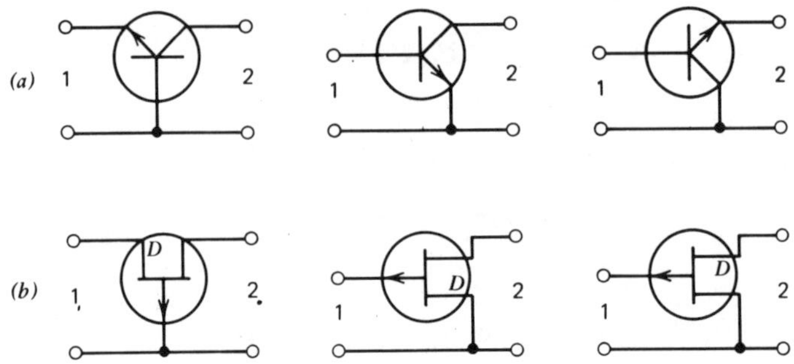

Figure 15.1. Schematic diagram of the three methods of connecting a transistor for (*a*) *bipolar transistor,* (*b*) *FET.*

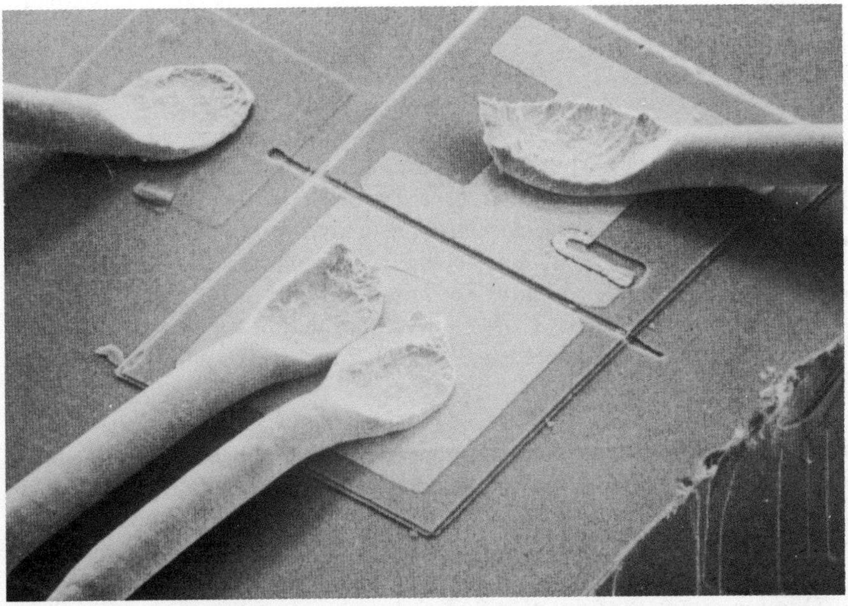

Figure 15.2. Plan view of a MESFET. (*Courtesey of the Plessey Company Ltd.*)

desirable to select such a transistor, since a small feedback factor means that the input characteristic of the amplifier is independent of the load, and the output characteristic is independent of the source conditions.

Figure 15.3 depicts a typical multistage amplifier. The diagram illustrates three types of amplifier design. A low noise figure first stage is used since the first stage noise figure determines the overall amplifier noise

Table 15.1 (after R.Niedert and H.A. Willing)

Freq. (GHz)	$\|S_{11}\| \underline{/\theta°}$	$\|S_{21}\| \underline{/\theta°}$	$\|S_{12}\| \underline{/\theta°}$	$\|S_{22}\| \underline{/\theta°}$
1.0	$0.970 \underline{/-46}$	$3.232 \underline{/144}$	$0.030 \underline{/61}$	$0.784 \underline{/-16}$
2.0	$0.886 \underline{/-85}$	$2.618 \underline{/113}$	$0.045 \underline{/35}$	$0.728 \underline{/-31}$
3.0	$0.874 \underline{/-113}$	$2.379 \underline{/90}$	$0.057 \underline{/18}$	$0.709 \underline{/-38}$
4.0	$0.821 \underline{/-155}$	$1.988 \underline{/59}$	$0.069 \underline{/-3}$	$0.681 \underline{/-51}$
5.0	$0.821 \underline{/-177}$	$1.567 \underline{/36}$	$0.064 \underline{/-22}$	$0.615 \underline{/-77}$
6.0	$0.815 \underline{/172}$	$1.383 \underline{/18}$	$0.064 \underline{/-28}$	$0.694 \underline{/-88}$
7.0	$0.812 \underline{/133}$	$1.348 \underline{/-8}$	$0.075 \underline{/-44}$	$0.750 \underline{/-92}$
8.0	$0.845 \underline{/111}$	$1.049 \underline{/-27}$	$0.063 \underline{/-70}$	$0.544 \underline{/-121}$
9.0	$0.894 \underline{/113}$	$0.899 \underline{/-53}$	$0.055 \underline{/-81}$	$0.694 \underline{/-154}$
10.0	$0.734 \underline{/106}$	$0.752 \underline{/-67}$	$0.059 \underline{/-92}$	$0.784 \underline{/-162}$
11.0	$0.810 \underline{/59}$	$0.850 \underline{/-91}$	$0.062 \underline{/-113}$	$0.594 \underline{/-169}$
12.0	$0.851 \underline{/45}$	$0.653 \underline{/-127}$	$0.058 \underline{/-139}$	$0.707 \underline{/146}$

Table 15.1 (after R.Niedert and H.A. Willing)

Freq. (GHz)	$\|S_{11}\| \underline{/\theta°}$	$\|S_{21}\| \underline{/\theta°}$	$\|S_{12}\| \underline{/\theta°}$	$\|S_{22}\| \underline{/\theta°}$
1.0	$0.968 \underline{/-42}$	$3.423 \underline{/145}$	$0.029 \underline{/62}$	$0.715 \underline{/-17}$
2.0	$0.916 \underline{/-84}$	$3.004 \underline{/112}$	$0.048 \underline{/34}$	$0.662 \underline{/-38}$
3.0	$0.858 \underline{/-114}$	$2.389 \underline{/85}$	$0.055 \underline{/9}$	$0.627 \underline{/-57}$
4.0	$0.845 \underline{/-133}$	$1.954 \underline{/64}$	$0.060 \underline{/-10}$	$0.631 \underline{/-70}$
5.0	$0.786 \underline{/-153}$	$1.721 \underline{/47}$	$0.057 \underline{/-34}$	$0.649 \underline{/-74}$
6.0	$0.770 \underline{/-177}$	$1.647 \underline{/28}$	$0.046 \underline{/-51}$	$0.662 \underline{/-76}$
7.0	$0.835 \underline{/151}$	$1.402 \underline{/6}$	$0.055 \underline{/-51}$	$0.627 \underline{/-91}$
8.0	$0.792 \underline{/130}$	$1.396 \underline{/-19}$	$0.073 \underline{/-69}$	$0.465 \underline{/-122}$
9.0	$0.774 \underline{/113}$	$1.217 \underline{/-46}$	$0.079 \underline{/-87}$	$0.501 \underline{/-155}$
10.0	$0.722 \underline{/96}$	$0.961 \underline{/-73}$	$0.085 \underline{/-105}$	$0.673 \underline{/-173}$
11.0	$0.652 \underline{/82}$	$0.673 \underline{/-86}$	$0.074 \underline{/-118}$	$0.746 \underline{/-178}$
12.0	$0.635 \underline{/77}$	$0.560 \underline{/-91}$	$0.064 \underline{/-119}$	$0.739 \underline{/-177}$

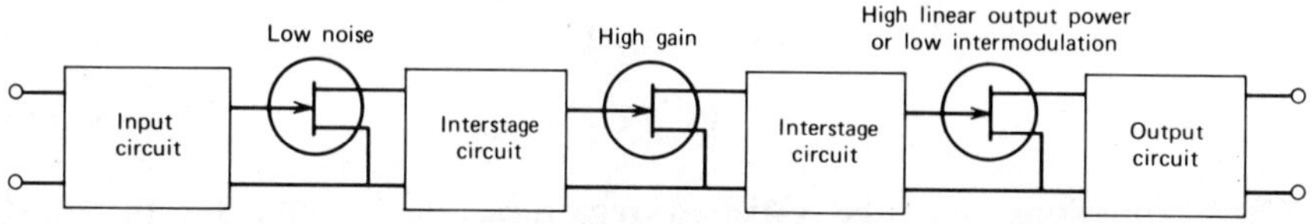

Figure 15.3. Block diagram of a typical multistage FET amplifier.

figure. The FET in the second stage is operated with maximum small-signal gain, and the last is a power or low intermodulation stage. The text is essentially devoted to the design of the maximum small-signal gain amplifier.

15.1 *STABILITY OF A 2-PORT NETWORK*

The primary consideration in the design of transistor amplifiers is to ensure that the circuit does not oscillate. An amplifier may exhibit conditional or unconditional stability. A circuit is unconditionally stable if its input and output resistances are positive for any passive terminations, this is equivalent to $|S'_{11}|$ and $|S'_{22}|$ below being less than unity. However, if $|S'_{11}|$ and $|S'_{22}|$ are greater than unity, the amplifier may still exhibit a conditional stability by restricting the elements of the matching network.

The input-output relation for the amplifier is

$$\begin{bmatrix} b_1 \\ b_2 \end{bmatrix} = \begin{bmatrix} S_{11} & S_{12} \\ S_{21} & S_{22} \end{bmatrix} \begin{bmatrix} a_1 \\ a_2 \end{bmatrix} \tag{15.1}$$

where

$$a_2 = \Gamma_L b_2 \tag{15.2}$$

when the network is terminated by a reflection coefficient Γ_L.

Substituting Eq. 15.2 into Eq. 15.1 gives

$$\frac{b_1}{a_1} = S'_{11} = S_{11} + \frac{S_{12}S_{21}\Gamma_L}{1-\Gamma_L S_{22}} \tag{15.3}$$

and by a similar procedure at the output port

$$\frac{b_2}{a_2} = S'_{22} = S_{22} + \frac{S_{21}S_{12}\Gamma_G}{1-\Gamma_G S_{11}} \tag{15.4}$$

These equations may also be obtained directly by setting $r_1 = 0$ in Eq. 1.112 and $r_2 = 0$ in Eq. 1.115.

For stability the boundary conditions on Eqs. 15.3 and 15.4 are

$$|S'_{11}| < 1 \tag{15.5}$$

$$|S'_{22}| < 1 \tag{15.6}$$

These equations must be satisfied regardless of Γ_L and Γ_G provided $|\Gamma_L|, |\Gamma_G| < 1$.

Equations 15.3 and 15.4 are standard bilinear transformations in complex variable theory. Under each transformation, a circle of constant Γ_L or Γ_G is tranformed into a circle in the S'_{11} or S'_{22} planes. This is a significant result, since it is now possible to define stability regions on the Smith chart representing the Γ_L or Γ_G planes.

Setting

$$|S'_{11}| = 1 \tag{15.7}$$

$$|S'_{22}| = 1 \tag{15.8}$$

and substituting into Eqs. 15.3 and 15.4 will give the limiting values of stability for Γ_L or Γ_G. The solutions lie on circles of center and radius

$$r_{S2} = \frac{C_2^*}{|S_{22}|^2 - |\Delta|^2} \tag{15.9}$$

$$\rho_{S2} = \frac{|S_{12}S_{21}|}{|S_{22}|^2 - |\Delta|^2} \tag{15.10}$$

on the Γ_L plane, and

$$r_{S1} = \frac{C_1^*}{|S_{11}|^2 - |\Delta|^2} \tag{15.11}$$

$$\rho_{S1} = \frac{|S_{12}S_{21}|}{|S_{11}|^2 - |\Delta|^2} \tag{15.12}$$

on the Γ_G plane, where

$$C_1 = S_{11} - \Delta S_{22}^* \tag{15.13}$$

$$C_2 = S_{22} - \Delta S_{11}^* \tag{15.14}$$

$$\Delta = S_{11}S_{22} - S_{12}S_{21} \tag{15.15}$$

A typical stability circle, illustrating the definition of $r_{S1,2}$ and $\rho_{S1,2}$ is depicted in Figure 15.4.

The stability circle defines the boundary between the stable and unstable regions on the Smith chart. This means that either the inside or the outside of the circle can represent the stable region. To determine which area represents the stable operating condition a suitable test point must be selected and checked for stability, using Eqs. 15.3 and 15.4 and the general stability conditions of Eqs. 15.7 and 15.8. The most suitable point is the

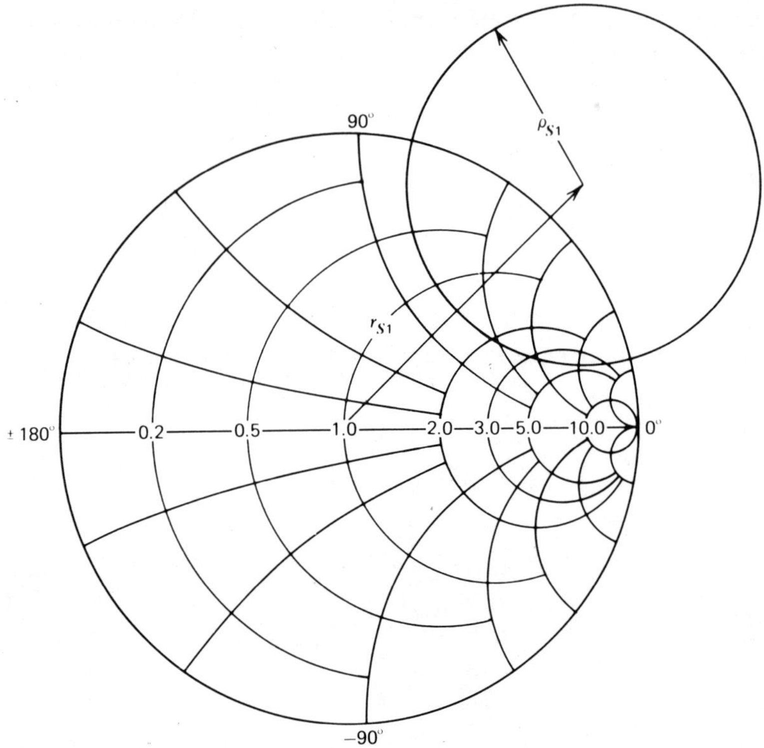

Figure 15.4. A typical stability circle on the Γ_L plane, illustrating the definition of r_{S2} and ρ_{S2}.

center of the Smith chart, i.e., the point $\Gamma_L = 0$. At this point Eq. 15.3 reduces to

$$S'_{11} = S_{11} \tag{15.16}$$

Recalling the conditions for stability gives

$$|S_{11}| < 1 \tag{15.17}$$

If this condition is satisfied, then the Smith chart center represents a stable region. The two possible cases of this condition are illustrated in Figures 15.5*a* and 15.5*b*. The first condition is where the stability circle encloses the origin and hence its interior represents the stable region. The second illustrates the case where the origin is outside the circle and the exterior of the circle represents the stable region.

If the stability conditions are not satisfied for the origin at $\Gamma_L = 0$, Eq.

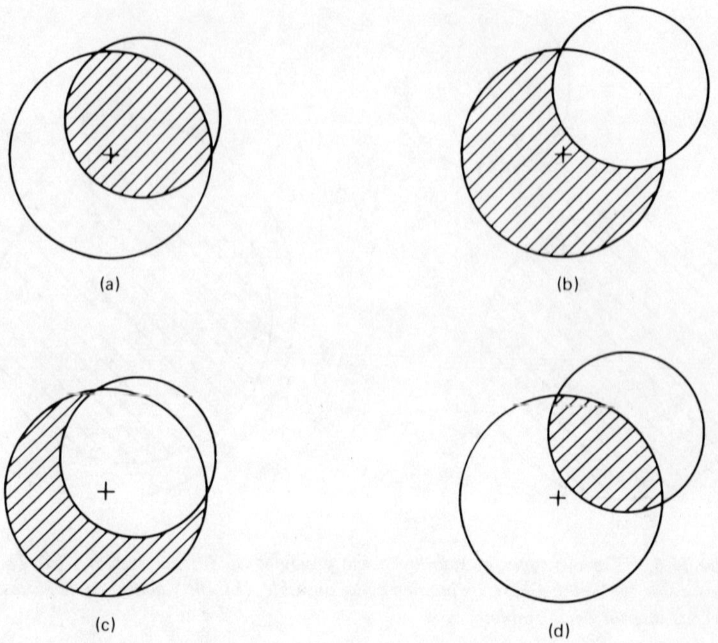

Figure 15.5. A schematic diagram of the four different regions of stability. (a), (b) The Smith Chart center is stable. (c), (d) The Smith Chart center is unstable.

15.17 becomes

$$|S_{11}| > 1$$

and the origin represents an unstable point. The two cases of this condition are shown in Figures 15.5*c* and 15.5*d*. Figure 15.5*c* is the case where the stability circle encircles the origin and its exterior is the stable region. The case where the circle does not encircle the origin, its interior representing the stable region is shown in Fig. 15.5*d*.

15.2 *UNCONDITIONAL STABILITY*

In this section, the conditions for an amplifier to be unconditionally stable will be derived. This can be interpreted as ensuring that the stable region defined by the stability circles extends throughout the whole Smith chart. The two types of stability circles illustrating a stable region on the inside and outside are shown in Figures 15.6*a* and 15.6*b*. The condition for the stable region at the output to extend throughout the complete Smith chart

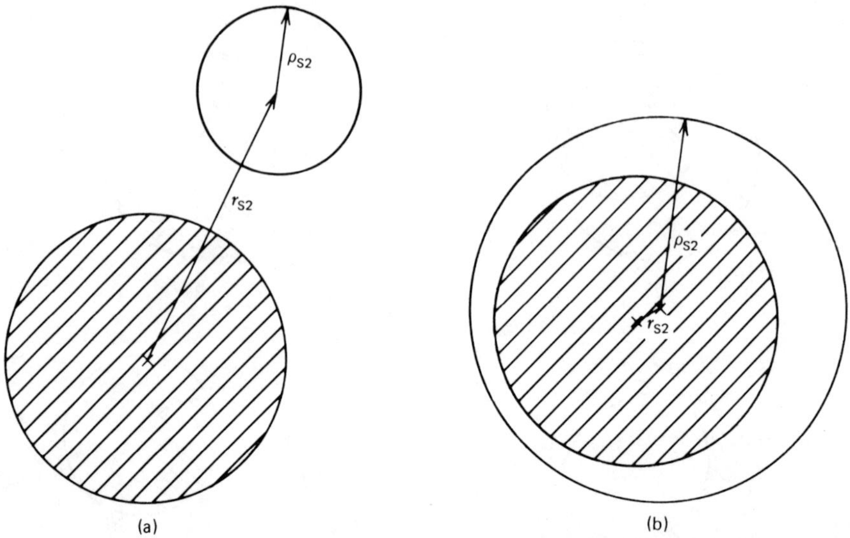

Figure 15.6. The two types of stability circle giving unconditional stability. (a) The stability circle outside the Smith Chart, its interior being unstable. (b) The stability circle enclosing Smith Chart, its interior being stable.

is

$$||r_{S2}| - \rho_{S2}| > 1 \tag{15.18}$$

In Figure 15.6*a* the limiting case of the above equation corresponds to letting

$$r_{S2} \to 1 \tag{15.19}$$

$$\rho_{S2} \to 0 \tag{15.20}$$

The unstable region in the limit is therefore reduced to a point on the Smith chart circumference.

In Figure 15.6*b* the limiting value of Eq. 15.18 corresponds to letting

$$r_{S2} \to 0 \tag{15.21}$$

$$\rho_{S2} \to 1 \tag{15.22}$$

In this instance the unstable region is the space immediately outside the Smith chart perimeter.

Substituting for r_{S2} and ρ_{S2} into Eq. 15.18 from Eqs. 15.9 and 15.10 gives

$$\left| \frac{|C_2^*| - |S_{12}S_{21}|}{|S_{22}|^2 - |\Delta|^2} \right| > 1 \qquad (15.23)$$

Interchanging the subscripts 1 and 2 gives the corresponding condition for the input

$$\left| \frac{|C_1^*| - |S_{21}S_{12}|}{|S_{11}|^2 - |\Delta|^2} \right| > 1 \qquad (15.24)$$

Eliminating C_2^* from Eqs. (15.23) with the help of Eq 15.14, the stability conditon of Eq. 15.18 becomes

$$2|S_{12}S_{21}| < 1 - |S_{11}|^2 - |S_{22}|^2 + |\Delta|^2 \qquad (15.25)$$

This equation is usually written in terms of a stability factor K as

$$K > 1 \qquad (15.26)$$

where

$$K = \frac{1 + |\Delta|^2 - |S_{11}|^2 - |S_{22}|^2}{2|S_{21}S_{12}|} \qquad (15.27)$$

This equation for K applies to both input and output circuits. Another relationship between the scattering parameters to ensure stability can be obtained by expanding Eq. (15.25) and rearranging to give

$$\left| S_{22} + \frac{S_{11}^* S_{12} S_{21}}{1 - |S_{11}|^2} \right| < 1 - \frac{|S_{12}S_{21}|}{1 - |S_{11}|^2} \qquad (15.28)$$

This is only true if

$$|S_{12}S_{21}| < 1 - |S_{11}|^2 \qquad (15.29)$$

Interchanging subscripts gives the condition for the input circuit

$$|S_{12}S_{21}| < 1 - |S_{22}|^2 \qquad (15.30)$$

Equations 15.27, 15.29, and 15.30 state the conditions, in terms of the transistor parameters, which must be satisfied to ensure unconditional stability.

15.3 DEFINITION OF TRANSDUCER POWER GAIN

In the design of small signal power amplifiers, a transducer power gain can be defined for the device terminated in arbitrary load and generator impedances.

The transducer power gain is defined as

$$G_T = \frac{\text{power delivered to load}}{\text{power available from source}} = |S'_{21}|^2$$

where $|S'_{21}|$ is given by Eq. 1.114 as

$$|S'_{21}|^2 = \frac{|S_{21}|^2(1-|\Gamma_G|^2)(1-|\Gamma_L|^2)}{|(1-S_{11}\Gamma_G)(1-S_{22}\Gamma_L)-S_{12}S_{21}\Gamma_L\Gamma_G|} \tag{15.31}$$

r_1 and r_2 have been replaced by Γ_G and Γ_L in Eqs. 1.112 through 1.115, with

$$\Gamma_G = \frac{Z_G - Z_0}{Z_G + Z_0} \tag{15.32}$$

$$\Gamma_L = \frac{Z_L - Z_0}{Z_L + Z_0} \tag{15.33}$$

and the scattering parameters have the standard meaning. The transducer power gain provides a measure of the advantage of using a transistor over matching the same load to the source.

When the scattering parameter S_{12} is sufficiently small to be completely neglected, the device is termed unilateral because there is no internal feedback. Setting $S_{12}=0$ in Eq. 15.31 defines the unilateral transducer power gain as

$$G_U = |S_{21}|^2 \cdot \frac{(1-|\Gamma_G|^2)}{|1-S_{11}\Gamma_G|^2} \cdot \frac{(1-|\Gamma_L|^2)}{|1-S_{22}\Gamma_L|^2} \tag{15.34}$$

The first term is related to the transistor being used and remains invariant throughout the design. The other two terms, however, are not only related to the remaining S parameters of the device S_{11} and S_{22} but also to the load and source reflection coefficients. These latter two quantities can be used to control the design of the amplifier.

The unilateral power gain can be decomposed into three separate independent gain blocks.

$$G_U = G_G \cdot G_0 \cdot G_L \tag{15.35}$$

where

$$G_G = \frac{1-|\Gamma_G|^2}{|1-S_{11}\Gamma_G|^2} \tag{15.36}$$

$$G_0 = |S_{21}|^2 \tag{15.37}$$

$$G_L = \frac{1-|\Gamma_L|^2}{|1-S_{22}\Gamma_L|^2} \tag{15.38}$$

Figure 15.7 depicts a block diagram of the amplifier subdivided into the three gain units; G_G, G_0, and G_L.

The G_G term affects the degree of mismatch between the source impedance and the input reflection coefficient of the 2-port and may provide a gain contribution greater than unity. This is true because the matching circuit can improve the match existing between the generator impedance and the input impedance of the two port network. The G_0 term is related to the transistor and its bias condition and is simply $|S_{21}|^2$. This gain term exhibits a 6-dB/octave gain reduction characteristic. The third term in the expression, G_L, serves the same function as the G_G term but affects the output rather than the input matching.

15.4 *COMPLEX CONJUGATE IMAGE MATCHING*

If the transistor is found to be unconditionally stable, it can be simultaneously complex conjugate image matched. This leads to the maximum power gain for the device.

For a conditionally stable transistor, however, matching networks must be designed with reference to the stability criterion already discussed.

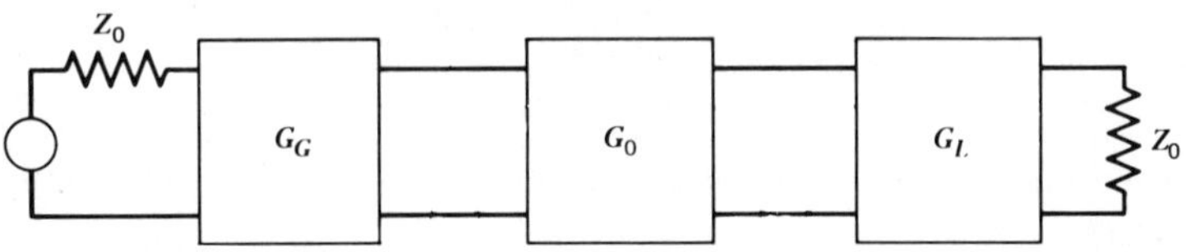

Figure 15.7. Block diagram of the unilateral amplifier illustrating the subdivision into three separate gain blocks.

Values of Γ_L and Γ_G can be determined by defining the stable operating regions on the Smith chart.

If an amplifier is unconditionally stable, the load and generator reflection coefficients to conjugate match the input and output ports are

$$\Gamma_G^* = S_{11} + \frac{S_{12}S_{21}\Gamma_L}{1-\Gamma_L S_{22}} \tag{15.39}$$

$$\Gamma_L^* = S_{22} + \frac{S_{12}S_{21}\Gamma_G}{1-\Gamma_G S_{11}} \tag{15.40}$$

Solving these two equations for Γ_L and Γ_G gives

$$\Gamma_G = C_1^* \frac{B_1 \pm \sqrt{B_1^2 - 4|C_1|^2}}{2|C_1|^2} \tag{15.41}$$

$$\Gamma_L = C_2^* \frac{B_2 \pm \sqrt{B_2^2 - 4|C_2|^2}}{2|C_2|^2} \tag{15.42}$$

where

$$B_1 = 1 + |S_{11}|^2 - |S_{22}|^2 - |\Delta|^2 \tag{15.43}$$

$$B_2 = 1 + |S_{22}|^2 - |S_{11}|^2 - |\Delta|^2 \tag{15.44}$$

and Δ, C_1, and C_2 are given by Eqs. 15.13 through 15.15.

If the computed value of B_1 or B_2 is negative, the plus sign should be used in Eqs. 15.41 and 15.42. Conversely, if B_1 or B_2 is positive the minus sign should be used. By taking the appropriate roots a value of less than unity is ensured for both reflection coefficients.

For complex conjugate image matching, the maximum transducer gain becomes from Eq. 15.31

$$G_{U,\max} = \left|\frac{S_{21}}{S_{12}}\right| \left(K \pm \sqrt{K^2 - 1}\right) \tag{15.45}$$

where the minus sign is used if B is positive and the plus sign is employed if B is negative.

In Eq. 15.45 K is given by Eq. 15.27.

For the unilateral amplifier, Eqs. 15.39 through 15.42 reduces to

$$\Gamma_G = S_{11}^* \tag{15.46}$$

$$\Gamma_L = S_{22}^* \tag{15.47}$$

The maximum unilateral power gain is, from Eq. 15.34

$$G_{U,\max} = \frac{1}{1-|S_{11}|^2} \cdot |S_{21}|^2 \cdot \frac{1}{1-|S_{22}|^2} \tag{15.48}$$

This has the form of Eq. 15.35

$$G_{U,\max} = G_{G,\max} \cdot G_0 \cdot G_{L,\max} \tag{15.49}$$

where $G_{G,\max}$ and $G_{L,\max}$ are the maximum values of the gain contributions due to the output and input matching networks.

For the unilateral amplifier the 2-port stability requirements reduce to 1-port stability conditions. For $|S_{11}|$ or $|S_{22}|$ less than unity the 1-port circuits are unconditionally stable, whereas for $|S_{11}|$ or $|S_{22}|$ larger than unity stability must be ensured by making the 1-port circuit positive real.

15.5 *CIRCLES OF CONSTANT UNILATERAL GAIN*

For conditions other than perfect conjugate matching, the unilateral power gain depends upon how G_L and G_G vary as the source and load impedances are changed.

Consider the expression for G_G in Eqs. 15.36. It has been shown that for complex conjugate image matching, i.e., $\Gamma_G = S_{11}^*$, G_G is a maximum. It is also clear that for $|\Gamma_G| = 1$, G_G has a value of zero. For any arbitrary constant value of G_G between these extremes of zero and $G_{G,\max}$, solutions for Γ_G lie on a circle.

The radius and the center of the constant gain circles can be obtained by first defining the normalized gain factors

$$g_G = \frac{G_G}{G_{G,\max}} = G_G\left(1-|S_{11}|^2\right) \tag{15.50}$$

$$g_L = \frac{G_L}{G_{L,\max}} = G_L\left(1-|S_{22}|^2\right) \tag{15.51}$$

where $G_{G,\max}$ and $G_{L,\max}$ are obtained from Eq. 15.49.

Γ_L and S_{22} can be expanded as

$$\Gamma_L = U + jV \tag{15.52}$$

$$S_{22} = A + jB \tag{15.53}$$

where U and jV represent the axes of the Smith chart. Substituting Eqs. 15.52 and 15.53 into Eq. 15.51 gives

$$\left[U-\frac{g_L A}{1-|S_{22}|^2(1-g_L)}\right]^2+\left[V+\frac{g_L B}{1-|S_{22}|^2(1-g_L)}\right]^2$$

$$=\left[\frac{\sqrt{1-g_L}\,(1-|S_{22}|^2)}{1-|S_{22}|^2(1-g_L)}\right]^2 \qquad (15.54)$$

This represents a family of circles in which the normalized power gain g_L is a parameter. The center and radius are given by

$$r_{0L}=\frac{g_L|S_{22}|^2}{1-|S_{22}|^2\,(1-g_L)} \qquad (15.55)$$

$$\rho_{0L}=\frac{(1-g_L)^{1/2}(1-|S_{22}|^2)}{1-|S_{22}|^2(1-g_L)} \qquad (15.56)$$

The center and radius of the constant gain circles for the input network are

$$r_{0G}=\frac{g_G|S_{11}|^2}{1-|S_{11}|^2\,(1-g_G)} \qquad (15.57)$$

$$\rho_{0G}=\frac{(1-g_G)^{1/2}(1-|S_{11}|^2)}{1-|S_{11}|^2(1-g_G)} \qquad (15.58)$$

Circles of constant gain g_G are illustrated in Figure 15.8. The circle that passes through the origin is always the unit gain circle for both G_L and G_G. In other words inside this circle both G_L and $G_G>1$ and outside G_L and $G_G<1$.

Constant gain circles can be constructed for $|S_{11}|$ or $|S_{22}|$ larger than unity in a similar manner to that used for the unconditionally stable situation by replacing S_{11} by $1/S_{11}$.

15.6 *UNILATERAL FIGURE OF MERIT*

Although the assumption has been made that S_{12} is sufficiently small as to be negligible in the amplifier design, it is desirable to determine how adequate the resultant unilateral design is.

The actual transducer power gain can be written in terms of the unilateral power gain from Eq. 15.31 as

$$G_T = G_U \cdot \frac{1}{|1-X|^2} \tag{15.59}$$

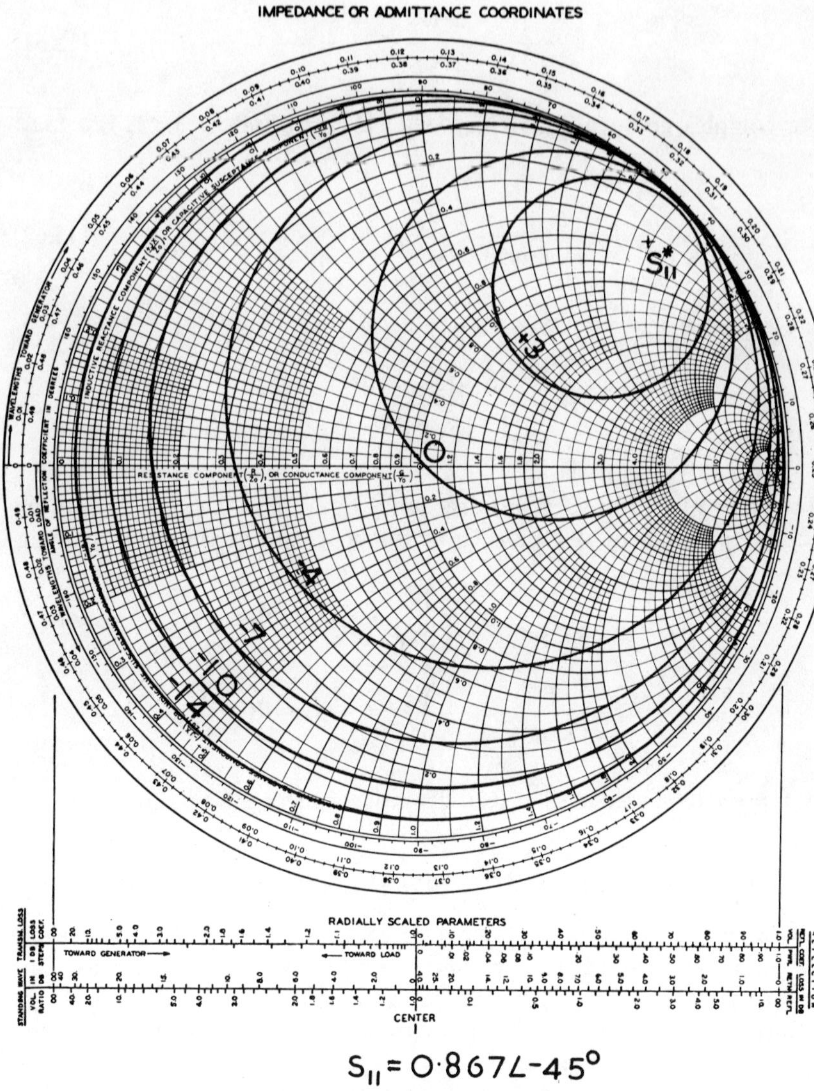

Figure 15.8. Circle of constant gain $g_{G'}$ plotted on the Smith Chart representing the Γ_G plane.

where

$$X = \frac{\Gamma_G \Gamma_L S_{12} S_{21}}{(1-\Gamma_G S_{11})(1-\Gamma_L S_{22})} \tag{15.60}$$

The ratio of the true gain to the unilateral gain is bounded by

$$\frac{1}{|1+|X||^2} < \frac{G_T}{G_U} < \frac{1}{|1-|X||^2} \tag{15.61}$$

For complex conjugate matching, with $|S_{11}| < 1$ and $|S_{22}| < 1$, Eq. 15.61 becomes

$$\frac{1}{(1+W)^2} < \frac{G_T}{G_U} < \frac{1}{(1-W)^2} \tag{15.62}$$

where

$$W = \frac{|S_{11}||S_{22}||S_{12}||S_{21}|}{(1-|S_{11}|^2)(1-|S_{22}|^2)} \tag{15.63}$$

The value of W will vary with frequency because of the variations of the scattering parameters. A typical value for W is 0.03, at 100 MHz yielding

$$\frac{1}{(1+W)^2} = \frac{1}{(1.03)^2} = -0.26 \text{ dB}$$

$$\frac{1}{(1-W)^2} = \frac{1}{(0.97)^2} = +0.26 \text{ dB}$$

This represents a spread of ± 0.26 dB in transducer power gain.

15.7 *EXAMPLE OF NARROW-BAND AMPLIFIER DESIGN*

This example deals with the synthesis of a simple amplifier, to operate between a source and load impedance of 50 Ω, which provides maximum gain at 750 MHz. The nonunilateral and unilateral cases will both be considered as an example of each design procedure. The scattering parameters of the transistor used are

$$S_{11} = 0.277 \angle -59.0 \text{ deg} \tag{15.64}$$

$$S_{12} = 0.078 \angle -93.0 \text{ deg}$$

$$S_{21} = 1.920 \angle \ 64.0 \text{ deg}$$

$$S_{22} = 0.848 \angle -31.0 \text{ deg}$$

15.8 *NONUNILATERAL DESIGN*

The scattering coefficients are now used to calculate

$$\Delta = S_{11}S_{22} - S_{12}S_{21} = 0.324 \angle -64.8 \text{ deg}$$

$$C_1 = S_{11} - \Delta S_{22}^* = 0.120 \angle -135.4 \text{ deg}$$

$$C_2 = S_{22} - \Delta S_{11}^* = 0.768 \angle -33.8 \text{ deg}$$

$$B_1 = 1 + |S_{11}|^2 - |S_{22}|^2 - |\Delta|^2 = 0.253$$

$$B_2 = 1 + |S_{22}|^2 - |S_{11}|^2 - |\Delta|^2 = 1.537$$

$$K = \frac{1 + |\Delta|^2 - |S_{11}|^2 - |S_{22}|^2}{2|S_{21}S_{21}|} = 1.033$$

By using the above values in Eqs. 15.26, 15.29, and 15.30, the transistor is unconditionally stable and complex conjugate matching can be used to realize the maximum transducer gain. B_1 and B_2 are both positive and therefore the negative signs must be used in Eqs. 15.41, 15.42 and 15.45 to yield

$$\Gamma_G = C_1^* \left[\frac{B_1 - \sqrt{B_1^2 - 4|C_1|^2}}{2|C_1|^2} \right] = 0.730 \angle 135.4 \text{ deg}$$

$$\Gamma_L = C_2^* \left[\frac{B_2 - \sqrt{B_2^2 - 4|C_2|^2}}{2|C_2|^2} \right] = 0.951 \angle 33.8 \text{ deg}$$

The values of impedance required to realize these reflection coefficients are

$$R_G = 9.1 + j19.9 \ \Omega$$

$$R_L = 14.7 + j163.1 \ \Omega$$

The gain is

$$G_{T,\max}=\left|\frac{S_{21}}{S_{12}}\right|(K-\sqrt{K^2-1})=19.087=12.807 \text{ dB}.$$

15.9 *UNILATERAL DESIGN*

In this section, the transistor is assumed unilateral, i.e.,

$$S_{12}\approx 0$$

The stability conditions in this case are

$$|S_{11}|<1$$

$$|S_{22}|<1$$

Using the scattering parameters in the example, the above are seen to be satisfied. Complex conjugate matching can be applied using Eqs. 15.46 and 15.47 to realize the maximum unilateral gain. This yields

$$\Gamma_G=S_{11}^*=0.277\angle 59.0 \text{ deg}$$

$$\Gamma_L=S_{22}^*=0.848\angle 32.0 \text{ deg}$$

The values of impedance required to realize these reflection coefficients are in this case

$$R_G=58+j30.0\ \Omega$$

$$R_L=52.5+j155.0\ \Omega$$

The unilateral amplifier gain is

$$G_{U,\max}=\frac{1}{1-|S_{11}|^2}\cdot|S_{21}|^2\cdot\frac{1}{1-|S_{22}|^2}=14.21=11.53 \text{ dB}.$$

The error in gain due to treating the transistor as unilateral is 1.28 db. This error can also be determined by using the unilateral figure of merit. Matching networks can be designed in both cases using standard network synthesis techniques.

PROBLEMS

1. Obtain the stability factor K at each frequency for the transistor characterized by Tables 15.1 and 15.2.
2. Verify that the maximum unilateral gain for the transistors in Tables 15.1 and 15.2 is approximately 6 dB/octave.
3. Calculate the maximum unilateral and non-unilateral gains at 2 GHz and 4 GHz for the transistor in Table 15.1. Verify that the difference in the two gains agrees with the unilateral figure of merit.
4. Draw stability circles at 2 and 4 GHz for the transistors in Tables 15.1 and 15.2.
5. Demonstrate how you would wire the 3-dB hybred to construct a balanced amplifier from two identical transistor gain stages.
6. Repeat the narrow-band design example in the text for the nonunilateral case to yield a gain of 10 dB.
7. Design an amplifier at 2 GHz using complex conjugate image matching for the transistor in Table 15.2.
8. Design an amplifier to operate over the frequency range 8–12 GHz to obtain the maximum flat gain response using the scattering parameters given in Table 15.1.
9. Calculate the error from Eq. 15.62 in assuming that the FET using the scattering variables in Table 15.2 is unilateral.

FURTHER READING

1. W. Froeher, Quick amplifier design with scattering parameters, *Electronics*, **40** (Oct. 1967), pp. 100–109.
2. F. Weinert, Scattering parameters speed design of high frequency transistor circuits, *Electronics*, **39** (1966), pp. 78–88.
3. G. E. Bodway, Circuit design and characterizations of transistors by means of three port scattering parameters, *Microwave J.*, **11**, 5 (May 1968).
4. G. E. Bodway, Two port power flow analysis using generalised scattering parameters, *Microwave J.*, **10**, 6 (May 1967).
5. R. W. Anderson, *S*-Parameter techniques for faster more accurate network design, *Hewlett-Packard J.*, **18**, 6 (Feb. 1967).
6. *S*-Parameter Design, Hewlett-Packard Application Note 154, April 1972.
8. K. Kurokawa, *Principles of Microwave Circuits*, Academic Press, New York, 1969.
9. P. J. Owens and D. Woods, Re-appraisal of the unconditional stability criteria for active 2-port networks in terms of *S*-parameters, *Electronic Lett.*, **6**, 10 (May 1970).
10. R.E. Niedert and H.A. Willing, Wideband Gallium Arsenide Power MESFET Amplifiers, *IEEE Trans. on MTT*, **24**, No. 6, (June 1976).

Index